GCSE
Physics

Complete Revision
and Practice

Contents

How Science Works

Theories Come, Theories Go 1
Your Data's Got to Be Good 3
Bias and How to Spot it 5
Science Has Limits 6

Section One — Heat and Energy *c*

Moving and Storing Heat 8
Melting and Boiling 10
    Warm-Up and Exam Questions 12
Conduction 13
Convection 14
Heat Radiation 15
Saving Energy 17
    Warm-Up and Exam Questions 18
Energy Transfer 19
Energy Transformation Diagrams 20
Efficiency 21
    Warm-Up and Exam Questions 23
Energy Sources 24
Nuclear Energy 26
Nuclear and Geothermal Energy 27
Wind and Solar Energy 28
Solar Energy 29
Biomass 30
Wave Energy 31
Hydroelectric Power 32
Pumped Storage and Power Stations 32
    Warm-Up and Exam Questions 34
    Exam Questions 35
Revision Summary for Section One 36

28 pages

Section Two — Electricity and Waves *b*

Electric Current 37
Current, Voltage and Resistance 38
The Dynamo Effect 40
Power Stations and the National Grid 42
    Warm-Up and Exam Questions 44
    Exam Questions 45
Electrical Power 46
    Warm-Up and Exam Questions 48
Waves—The Basics 49
Electromagnetic Waves 50
Wave Speed and Interference 51
Diffraction 52
Refraction 53
    Warm-Up and Exam Questions 55
    Exam Questions 56
Dangers of EM Radiation 57
    Warm-Up and Exam Questions 59
Uses of Waves 60
    Warm-Up and Exam Questions 64
Analogue and Digital Signals 65
Seismic Waves 67
    Warm-Up and Exam Questions 69
Revision Summary for Section Two 70

33 pages

Section Three — Radioactivity and Space *e*

Radioactivity 71
Background Radiation 74
Half-Life 76
    Warm-Up and Exam Questions 78
Dangers from Nuclear Radiation 79
Uses of Nuclear Radiation 81
Radioactive Dating 83
    Warm-Up and Exam Questions 85
The Solar System 86
Magnetic Fields and Solar Flares 88
Beyond the Solar System 90
    Warm-Up and Exam Questions 91
The Life Cycle of Stars 92
The Origins of the Universe 93
    Warm-Up and Exam Questions 95
Exploring the Solar System 96
Looking into Space 98
    Warm-Up and Exam Questions 100
Revision Summary for Section Three 101

30

Section Four — Forces and Energy *e*

Velocity and Acceleration 102
D-T and V-T Graphs 104
Mass, Weight and Gravity 106
    Warm-Up and Exam Questions 107
    Exam Questions 108
The Three Laws of Motion 109
Friction Forces 112
Terminal Speed 113
    Warm-Up and Exam Questions 114
Stopping Distances 115
Momentum and Collisions 116
Car Safety 118
    Warm-Up and Exam Questions 120
Work and Potential Energy 121
Kinetic Energy 124
Roller Coasters 125
    Warm-Up and Exam Questions 126
Power 127
Fuels for Cars 129
    Warm-Up and Exam Questions 130
Revision Summary for Section Four 131

29

Section Five — Electricity *M*

Static Electricity 132
Uses of Static Electricity 136
    Warm-Up and Exam Questions 138
    Exam Questions 139
Circuits — the Basics 140
Resistance and Devices 141
Measuring AC 142
Series Circuits 144
Parallel Circuits 145
    Warm-Up and Exam Questions 146
Fuses and Safe Plugs 147
Energy and Power in Circuits 149
Charge, Voltage and Energy 151
    Warm-Up and Exam Questions 152
Revision Summary for Section Five 153

21

Contents

Section Six — Mechanics

Relative Speed and Velocity 154
Combining Velocities and Forces 155
Equations of Motion 157
Projectile Motion 159
    Warm-Up and Exam Questions 160
Turning Forces and Centre of Mass 161
Balanced Moments and Stability 163
    Warm-Up and Exam Questions 165
Circular Motion 166
Satellites 167
Gravity and Orbits 169
    Warm-Up and Exam Questions 170
Revision Summary for Section Six 171

Section Seven — Generating Electricity

Magnetic Fields 172
The Motor Effect 174
The Simple Electric Motor 176
    Warm-Up and Exam Questions 178
Generators 179
Transformers 180
    Warm-Up and Exam Questions 183
Nuclear Power 184
Nuclear Fusion 186
    Warm-Up and Exam Questions 187
Revision Summary for Section Seven 188

Section Eight — Wave Behaviour

Images 189
Mirrors 191
    Warm-Up and Exam Questions 195
Refractive Index and Snell's Law 196
    Warm-Up and Exam Questions 199
Lenses 200
Converging Lenses 202
Diverging Lenses 203
Uses — Magnification and Cameras 204
    Warm-Up and Exam Questions 205
Interference of Waves 206
Diffraction Patterns and Polarisation 208
    Warm-Up and Exam Questions 209
Sound Waves 210
Ultrasound 213
    Warm-Up and Exam Questions 217
Revision Summary for Section Eight 218

Section Nine — Circuits and Logic Gates

Potential Dividers 219
Diodes and Rectification 221
Capacitors 223
    Warm-Up and Exam Questions 225
Logic Gates 226
Using Logic Gates 228
LEDs and Relays in Logic Circuits 230
    Warm-Up and Exam Questions 231
Revision Summary for Section Nine 232

Section Ten — Particles in Action

Kinetic Theory and Temperature in Gases 233
Kinetic Theory and Pressure in Gases 235
    Warm-Up and Exam Questions 237
Particles in Atoms 238
Fundamental and Other Particles 241
    Warm-Up and Exam Questions 242
Electron Beams 243
    Warm-Up and Exam Questions 246
Revision Summary for Section Ten 247

Section Eleven — Medical Physics

Medical Uses of Light 248
Energy and Metabolic Rate 250
Electricity and the Body 251
    Warm-Up and Exam Questions 253
Intensity of Radiation 254
Nuclear Bombardment 256
    Warm-Up and Exam Questions 258
Momentum Conservation 259
Medical Uses of Radiation 261
Medical Research 263
    Warm-Up and Exam Questions 264
Revision Summary for Section Eleven 265

Exam Skills

Answering Experiment Questions 266

Answers 270

Index 282

Published by Coordination Group Publications Ltd

Editors:
Amy Boutal, Ellen Bowness, Mary Falkner, Gemma Hallam, Tom Harte, Sarah Hilton,
Kate Houghton, Paul Jordin, Sharon Keeley, Sam Norman, Andy Park, Kate Redmond,
Alan Rix, Ami Snelling, Rachel Selway, Laurence Stamford, Claire Thompson,
Jennifer Underwood, Julie Wakeling, Sarah Williams.

Contributors:
Mark A. Edwards, Sandy Gardner, Judith Hayes, Jason Howell, Barbara Mascetti,
John Myers, Richard Parsons, Moira Steven, Luke Waller, Andy Williams.

ISBN: 978 1 84146 657 6

With thanks to Mark A. Edwards and Glenn Rogers for the proofreading.
With thanks to Laura Phillips for the copyright research.

With thanks to Science Photo Library for permission to reproduce the photographs
used on pages 17, 89, 98 and 99.

Data used to construct stopping distance diagram on page 115 from the Highway Code.
Reproduced under the terms of the Click-Use license.

Groovy website: www.cgpbooks.co.uk

Printed by Elanders Hindson Ltd, Newcastle upon Tyne.
Jolly bits of clipart from CorelDRAW®

Theories Come, Theories Go

SCIENTISTS ARE ALWAYS RIGHT — OR ARE THEY?

Well it'd be nice if that were so, but it just ain't — never has been and never will be.
Increasing scientific knowledge involves making mistakes along the way. Let me explain...

Scientists come up with **hypotheses** — then **test** them

1) Scientists try and <u>explain</u> things. Everything.

2) They start by <u>observing</u> or <u>thinking about</u> something they don't understand — it could be anything,
e.g. planets in the sky, a person suffering from an illness, what matter is made of... anything.

3) Then, using what they already know (plus a bit of insight),
they come up with a <u>hypothesis</u> (a <u>theory</u>) that could <u>explain</u>
what they've observed.

*About 500 years ago, we
thought the Solar System
looked like this.*

> Remember, a hypothesis is just a <u>theory</u>, a
> <u>belief</u>. And <u>believing</u> something is true doesn't
> <u>make</u> it true — not even if you're a scientist.

4) So the next step is to try and convince other scientists that
the hypothesis is right — which involves using <u>evidence</u>.
First, the hypothesis has to fit the <u>evidence</u> already available
— if it doesn't, it'll convince <u>no one</u>.

5) Next, the scientist might use the hypothesis to make a <u>prediction</u> — a crucial step. If the hypothesis
predicts something, and then <u>evidence</u> from <u>experiments</u> backs that up, that's pretty convincing.

> This <u>doesn't</u> mean the hypothesis is <u>true</u> (the 2nd prediction, or the
> 3rd, 4th or 25th one might turn out to be <u>wrong</u>) — but a hypothesis
> that correctly predicts something in the <u>future</u> deserves respect.

A hypothesis is a good place to start
You might have thought that science was all about facts... well, it's not as cut and dried as
that — you also need to know about the process that theories go through to become accepted,
and how those theories change over time. Remember, nothing is set in stone...

Theories Come, Theories Go

Other scientists will **test** the hypotheses too

1) Now then... <u>other</u> scientists will want to use the hypothesis to make their <u>own predictions</u>, and they'll carry out their <u>own experiments</u>. (They'll also try to <u>reproduce</u> earlier results.) And if all the experiments in all the world back up the hypothesis, then scientists start to have a lot of <u>faith</u> in it.

2) However, if a scientist somewhere in the world does an experiment that <u>doesn't</u> fit with the hypothesis (and other scientists can <u>reproduce</u> these results), then the hypothesis is in trouble. When this happens, scientists have to come up with a new hypothesis (maybe a <u>modification</u> of the old theory, or maybe a completely <u>new</u> one).

3) This process of testing a hypothesis to destruction is a vital part of the scientific process. Without the '<u>healthy scepticism</u>' of scientists everywhere, we'd still believe the first theories that people came up with — like thunder being the belchings of an angered god (or whatever).

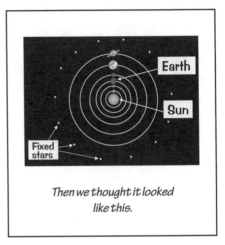

Then we thought it looked like this.

If **evidence** supports a hypothesis, it's **accepted** — for now

1) If pretty much every scientist in the world believes a hypothesis to be true because experiments back it up, then it usually goes in the <u>textbooks</u> for students to learn.

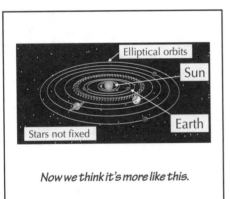

Now we think it's more like this.

2) Our <u>currently accepted</u> theories are the ones that have survived this 'trial by evidence' — they've been tested many, many times over the years and survived (while the less good ones have been ditched).

3) However... they never, <u>never</u> become hard and fast, totally indisputable <u>fact</u>.

> You can never know... it'd only take <u>one</u> odd, totally inexplicable result, and the hypothesising and testing would start all over again.

You expect me to believe that — then show me the evidence...

If scientists think something is true, they need to produce evidence to convince others — it's all part of <u>testing a hypothesis</u>. One hypothesis might survive these tests, while others won't — it's how things progress. And along the way some hypotheses will be disproved — i.e. shown not to be true. So, you see... not everything scientists say is true. <u>It's how science works</u>.

Your Data's Got to Be Good

Evidence is the key to science — but not all evidence is equally good.
The way that evidence is gathered can have a big effect on how trustworthy it is.

Lab experiments are better than rumour or small samples

1) Results from controlled experiments in laboratories are great. A lab is the easiest place to control variables so that they're all kept constant (except for the one you're investigating).

   This makes it easier to carry out a fair test.

   It's also the easiest way for different scientists around the world to carry out the same experiments. (There are things you can't study in a lab though, like climate.)

2) Old wives' tales, rumours, hearsay, 'what someone said', and so on, should be taken with a pinch of salt. They'd need to be tested in controlled conditions to be genuinely scientific.

3) Data based on samples that are too small don't have much more credibility than rumours do.

   A sample should be representative of the whole population (i.e. it should share as many of the various characteristics in the population as possible) — a small sample just can't do that.

Evidence is only reliable if other people can repeat it

Scientific evidence needs to be reliable (or reproducible). If it isn't, then it doesn't really help.

> **RELIABLE** means that the data can be reproduced by others.

Example: Cold Fusion

In 1989, two scientists claimed that they'd produced 'cold fusion' (the energy source of the Sun — but without the enormous temperatures).

It was huge news — if true, this could have meant energy from sea water — the ideal energy solution for the world... forever.

However, other scientists just couldn't get the same results — i.e. the results weren't reliable. And until they are, 'cold fusion' isn't going to be generally accepted as fact.

Reliability is really important in science

The scientific community won't accept someone's data if it can't be repeated by anyone else. It may sound like a really fantastic new theory, but if there's no other support for it, it just isn't reliable.

Your Data's Got to Be Good

Evidence also needs to be *valid*

To answer scientific questions scientists often try to <u>link</u> changes in <u>one</u> variable with changes in <u>another</u>. This is useful evidence, as long as it's <u>valid</u>.

> <u>**VALID**</u> means that the data is <u>reliable</u> AND <u>answers the original question</u>.

Example: Do power lines cause cancer?

Some studies have found that children who live near <u>overhead power lines</u> are more likely to develop <u>cancer</u>. What they'd actually found was a <u>correlation</u> between the variables "<u>presence of power lines</u>" and "<u>incidence of cancer</u>" — they found that as one changed, so did the other.

But this evidence is <u>not enough</u> to say that the power lines <u>cause</u> cancer, as other explanations might be possible.

For example, power lines are often near <u>busy roads</u>, so the areas tested could contain <u>different levels</u> of <u>pollution</u> from traffic. Also, you need to look at types of neighbourhoods and <u>lifestyles</u> of people living in the tested areas (could diet be a factor... or something else you hadn't thought of).

So these studies don't show a definite link and so don't <u>answer the original question</u>.

Controlling all the variables is *really hard*

In reality, it's <u>very hard</u> to control <u>all the variables</u> that might (just might) be having an effect.

You can do things to help — e.g. <u>choose</u> two <u>groups</u> of people (those near power lines and those far away) who are <u>as similar as possible</u> (same mix of ages, same mix of diets etc). But you can't easily rule out every possibility.

If you could do a <u>properly controlled lab experiment</u>, that'd be better — but you just can't do it without cloning people and exposing them to things that might cause cancer... <u>hardly ethical</u>.

Does the data really say that?

If it's so hard to be <u>definite</u> about anything, how does anybody ever get convinced about anything? Well, what usually happens is that you get a <u>load</u> of evidence that all points the same way. If one study can't rule out a particular possibility, then maybe another one can. So you gradually build up a whole <u>body of evidence</u>, and it's this (rather than any single study) that <u>convinces people</u>.

Bias and How to Spot it

Scientific results are often used to make a point, but results are sometimes presented in a biased way.

You don't need to *lie* to make things *biased*

1) For something to be misleading, it doesn't have to be untrue. We tend to read scientific facts and assume that they're the 'truth', but there are many different sides to the truth. Look at this headline...

> ### 1 in 2 people are of above average weight

 Sounds like we're a nation of fatties.

2) But an average is a kind of 'middle value' of all your data. Some readings are higher than average (about half of them, usually). Others will be lower than average (the other half).

So the above headline could just as accurately say:

> ### 1 in 2 people are of below average weight

3) The point is... both headlines sound quite worrying, even though they're not. That's the thing... you can easily make something sound really good or really bad — even if it isn't. You can...

---

① ...use only some of the data, rather than all of it:

> "Many people lost weight using the new SlimAway diet. Buy it now!!"

"Many" could mean anything — e.g. 50 out of 5000 (i.e. 1%). But that could be ignoring most of the data.

② ...phrase things in a 'leading' way:

> 90% fat free!

Would you buy it if it were "90% cyanide free"? That 10% is the important bit, probably.

③ ...use a statistic that supports your point of view:

> The amount of energy wasted is increasing.

> Energy wasted per person is decreasing.

> The rate at which energy waste is increasing is slowing down.

These describe the same data. But two sound positive and one negative.

Think about *why* things *might* be *biased*

1) People who want to make a point can sometimes present data in a biased way to suit their own purposes (sometimes without knowing they're doing it).

2) And there are all sorts of reasons why people might want to do this — for example...

- Governments might want to persuade voters, other governments, journalists, etc. Evidence might be ignored if it could create political problems, or emphasised if it helps their cause.
- Companies might want to 'big up' their products. Or make impressive safety claims, maybe.
- Environmental campaigners might want to persuade people to behave differently.

3) People do it all the time. This is why any scientific evidence has to be looked at carefully. Are there any reasons for thinking the evidence is biased in some way?

- Does the experimenter (or the person writing about it) stand to gain (or lose) anything?
- Might someone have ignored some of the data for political or commercial reasons?
- Is someone using their reputation rather than evidence to help make their case?

Scientific data's not always misleading, you just need to be careful. The most credible argument will be the one that describes all the data that was found, and gives the most balanced view of it.

Science Has Limits

Science can give us amazing things — cures for diseases, space travel, heated toilet seats...
But science has its limitations — there are questions that it just can't answer.

Some questions are **unanswered** by science — so far

1) We don't understand everything. And we never will. We'll find out more, for sure
— as more hypotheses are suggested, and more experiments are done.
But there'll always be stuff we don't know.

> For example, today we don't know as much as we'd like about
> climate change (global warming). Is climate change definitely
> happening? And to what extent is it caused by humans?

2) These are complicated questions, and at the moment scientists don't all agree on the answers.
But eventually, we probably will be able to answer these questions once and for all.

3) But by then there'll be loads of new questions to answer.

Other questions are **unanswerable** by science

1) Then there's the other type... questions that all the experiments in the world won't
help us answer — the "Should we be doing this at all?" type questions.
There are always two sides...

> The question of whether something is morally or ethically right
> or wrong can't be answered by more experiments — there is
> no "right" or "wrong" answer.

2) The best we can do is get a consensus from society — a judgement that most people are
more or less happy to live by. Science can provide more information to help people
make this judgement, and the judgement might change over time. But in the end it's up
to people and their conscience.

To answer or not to answer, that is the question...

It's official — no one knows everything. Your teacher/mum/annoying older sister (delete as applicable)
might think and act as if they know it all, but sadly they don't. So in reality you know one thing they
don't — which clearly makes you more intelligent and generally far superior in every way. Possibly.

Science Has Limits

People have *different opinions* about *ethical questions*

1) Take the <u>atom bomb</u>. After hearing the <u>Nazis</u> were trying to build a nuclear bomb, the <u>Allies</u> began their own research, and got there first. The whole thing was <u>top secret</u> — but even amongst the people in the know, there were questions over whether they should use it...

- President Truman said it was <u>necessary</u>... it could <u>force Japan to surrender</u>, and <u>end the war</u> quickly. If they didn't use the bombs, then the <u>lives of countless American soldiers</u> might be lost in the continuing war with Japan. And the targets were justifiable as they were the sites of <u>military bases</u>.

- Some scientists said they should <u>not be used</u>... they said that there would be <u>no limit</u> to the <u>destructive power</u> of this new development, and that using the bombs in warfare would "open the door to an <u>era of devastation</u> on an <u>unimaginable scale</u>".

2) As we know, the bombs <u>were used</u> — that can't be changed. But the original discussions are still <u>relevant</u>. Should we develop <u>more powerful</u> nuclear bombs? Right now, <u>some countries</u> think not, and have signed a <u>treaty</u> not to. But <u>other countries</u> may have reason to think otherwise.

Loads of other *factors* can *influence decisions* too

Here are some other factors that can influence decisions about science, and the way science is used:

Economic factors

- <u>Companies</u> very often won't pay for research unless there's likely to be a <u>profit</u> in it.

- Society can't always <u>afford</u> to do things scientists recommend without <u>cutting back elsewhere</u> (e.g. investing heavily in alternative energy sources).

Social factors

- Decisions based on scientific evidence affect <u>people</u> — e.g. should fossil fuels be taxed more highly (to invest in alternative energy)? Should alcohol be banned (to prevent health problems)? <u>Would the effect on people's lifestyles be acceptable...</u>

Environmental factors

- Genetically modified crops may help us produce more food — but some people say they could cause <u>environmental problems</u>.

Science is a "real-world" subject...
Science isn't just done by people in white coats in labs who have no effect on the outside world. Science has a massive effect on the real world every day, and so real-life things like <u>money</u>, <u>morals</u> and <u>how people might react</u> need to be considered. It's why a lot of issues are so difficult to solve.

Moving and Storing Heat

When it starts to get a bit nippy, on goes the heating to warm things up a bit. Heating is all about the <u>transfer of energy</u>. Here are a few useful definitions to begin with.

Heat is a measure of energy

1) When a substance is <u>heated</u>, its particles gain <u>energy</u>. This energy makes the particles in a <u>gas or a liquid</u> move around <u>faster</u>. In a solid, the particles <u>vibrate more rapidly</u>.

2) This energy is measured on an <u>absolute scale</u>. (This means it can't go lower than zero, because there's a limit to how slow particles can move.) The unit of heat energy is the <u>joule (J)</u>.

Temperature is a measure of hotness

1) The <u>hotter</u> something is, the <u>higher</u> its <u>temperature</u>.

2) Temperature is usually measured in <u>°C</u> (degrees Celsius), but there are other temperature scales, like <u>°F</u> (degrees Fahrenheit).

<u>Energy</u> tends to <u>flow</u> from <u>hot objects</u> to <u>cooler</u> ones — e.g. warm radiators heat the cold air in your room. And the <u>bigger the temperature difference</u>, the <u>faster heat is transferred</u>. Kinda makes sense.

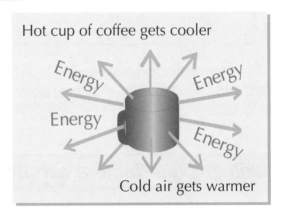

Hot cup of coffee gets cooler

Energy   Energy   Energy   Energy

Cold air gets warmer

If there's a <u>DIFFERENCE IN TEMPERATURE</u> between two places, then <u>ENERGY WILL FLOW</u> between them.

Specific heat capacity tells you how much energy stuff can store

1) It takes more <u>heat energy</u> to <u>increase</u> the <u>temperature</u> of some <u>materials</u> than others.

2) Materials which need to <u>gain</u> lots of energy to <u>warm up</u> also <u>release</u> loads of energy when they <u>cool down</u> again. They can '<u>store</u>' a lot of heat.

E.g. you need <u>4200 J</u> to warm 1 kg of <u>water</u> by 1 °C, but only <u>139 J</u> to warm 1 kg of <u>mercury</u> by 1 °C.

3) The measure of <u>how much energy</u> a substance can <u>store</u> is called its <u>specific heat capacity</u>.

It's all about energy...

Heat is energy, heat is energy — get that in your head and this topic will really start to make sense. It also explains why heat flows from hot to cold. Particles are lazy — they don't want to be buzzing around with lots of energy, they'd much rather pass that energy on and take it easy. Chill...

Moving and Storing Heat

Specific heat capacity is different for different materials — which makes it a favourite calculation topic for examiners. Make sure you know the formula below so they don't catch you out.

You need to be able to calculate specific heat capacity

1) Specific heat capacity is the amount of energy needed to raise the temperature of 1 kg of a substance by 1 °C. Water has a specific heat capacity of 4200 J/kg/°C.

2) The specific heat capacity of water is high. Once water's heated, it stores a lot of energy, which makes it good for central heating systems. Also, water's a liquid so it can easily be pumped around a building.

3) You'll have to do calculations involving specific heat capacity. This is the equation to learn:

Energy = Mass × Specific Heat Capacity × Temperature Change

EXAMPLE: How much energy is needed to heat 2 kg of water from 10 °C to 100 °C?

ANSWER: Energy needed = 2 × 4200 × 90 = 756 000 J

If you're not working out the energy, you'll have to rearrange the equation, so this formula triangle will come in dead handy.

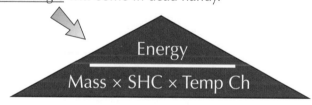

$$\frac{\text{Energy}}{\text{Mass} \times \text{SHC} \times \text{Temp Ch}}$$

You cover up the thing you're trying to find.
The parts of the formula you can still see are what it's equal to.

EXAMPLE: An empty 200 g aluminium kettle cools down from 115 °C to 10 °C, losing 19 068 J of heat energy. What is the specific heat capacity of aluminium?

ANSWER: $\text{SHC} = \dfrac{\text{Energy}}{\text{Mass} \times \text{Temp Ch}} = \dfrac{19\ 068}{0.2 \times 105} = \underline{908 \text{ J/kg/°C}}$

Remember — you need to convert the mass to kilograms first.

I wish I had a high specific fact capacity...

So there are two reasons why water's used in central heating systems — it's a liquid and it has a high specific heat capacity. This makes water good for cooling systems too. Water can absorb a lot of energy and carry it away. Water-based cooling systems are used in car engines and some computers.

Melting and Boiling

If you've ever made a cup of tea you'll know you need energy to boil water — usually supplied by the kettle. But the energy doesn't just go into raising the water temperature...

You need to put in energy to break intermolecular bonds

1) When you heat a liquid, the <u>heat energy</u> makes the <u>particles move faster</u>. Eventually, when enough of the particles have enough energy to overcome their attraction to each other, big bubbles of <u>gas</u> form in the liquid — this is <u>boiling</u>.

2) It's similar when you heat a solid. <u>Heat energy</u> makes the <u>particles vibrate faster</u> until eventually the forces between them are overcome and the particles start to move around — this is <u>melting</u>.

Heating

When a substance is <u>melting</u> or <u>boiling</u>, you're still putting in <u>energy</u>, but the energy's used for <u>breaking intermolecular bonds</u> rather than raising the temperature — there are <u>flat spots</u> on the heating graph.

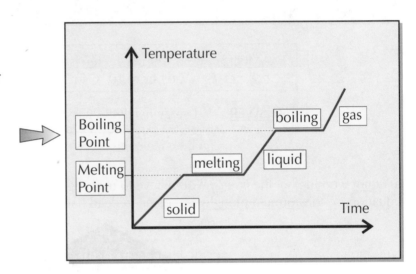

Cooling

When a substance is <u>condensing</u> or <u>freezing</u>, bonds are <u>forming</u> between particles, which <u>releases</u> energy. This means the <u>temperature doesn't go down</u> until all the substance has turned into a liquid (condensing) or a solid (freezing).

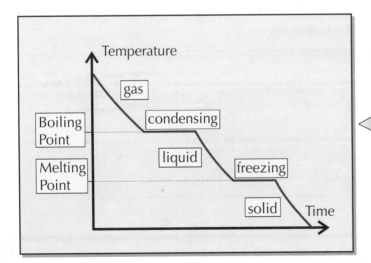

Melting and Boiling

If you heat up a pan of water on the stove, the water never gets any hotter than 100 °C. You can <u>carry on heating it up</u>, but the <u>temperature won't rise</u>. It's all to do with <u>latent heat</u>...

Specific latent heat is the energy needed to change state

1) The <u>specific latent heat of melting</u> is the <u>amount of energy</u> needed to <u>melt 1 kg</u> of material <u>without changing its temperature</u> (i.e. the material's got to be at its melting temperature already).

2) The <u>specific latent heat of boiling</u> is the <u>energy</u> needed to <u>boil 1 kg</u> of material <u>without changing its temperature</u> (i.e. the material's got to be at its boiling temperature already).

3) Specific latent heat is <u>different</u> for <u>different materials</u>, and it's different for <u>boiling</u> and <u>melting</u>. You don't have to remember what all the numbers are, though. Phew.

4) There's a <u>formula</u> to help you with all the <u>calculations</u>. And here it is:

Energy = Mass × Specific Latent Heat

EXAMPLE: The specific latent heat of water (for melting) is 334 000 J/kg. How much energy is needed to melt an ice cube of mass 7 g at 0 °C?

ANSWER: Energy = 0.007 × 334 000 J = <u>2338 J</u>

If you're finding the mass or the specific latent heat you'll need to divide, not multiply — just to make your life a bit easier, here's the formula triangle.

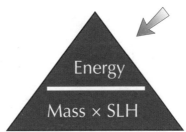

EXAMPLE: The specific latent heat of water (for boiling) is 2 260 000 J/kg. 2 825 000 J of energy is used to boil dry a pan of water at 100 °C. What was the mass of water in the pan?

ANSWER: Mass = Energy ÷ SLH = 2 825 000 ÷ 2 260 000 J = <u>1.25 kg</u>

It's all quite complicated but you really need to learn it

Melting a solid or boiling a liquid means you've got to <u>break bonds</u> between particles. That takes energy. Specific latent heat is just the amount of energy you need per kilogram of stuff. Incidentally, this is how <u>sweating</u> cools you down — your body heat's used to change liquid sweat into gas. Nice.

Warm-Up and Exam Questions

Take your time with these questions — and don't miss out the tricky-looking parts.
If any of them baffle you, it's not too late to take another peek over the section.

Warm-Up Questions

1) What is heat?
2) Under what conditions will heat flow?
3) When a boiling liquid is heated, what is the heat energy used for?
4) Explain what is meant by 'specific latent heat of melting'.

Exam Questions

1   A 0.5 kg iron block is heated from 20 °C to 100 °C. This takes 18 000 J of energy.
    The specific heat capacity of iron is

   A   250 J/kg°C

   B   450 J/kg°C

   C   650 J/kg°C

   D   850 J/kg°C

   (1 mark)

2   The graph shows a cooling curve for a gas. Match the labels **A**, **B**, **C** and **D** with
    points **1- 4** on the graph.

   A   Cooling liquid

   B   Condensing

   C   Freezing

   D   Cooling gas

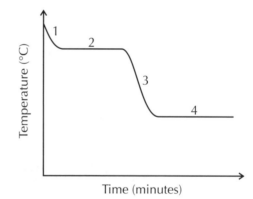

   (4 marks)

3   Sam is boiling some water. She uses a kettle which delivers 2000 J every second.
    Assume the kettle is 100% efficient.

   (a) How much heat is delivered to the water in two minutes?

   (2 marks)

   (b) The specific latent heat for boiling water is 2 260 000 J/kg. Once the water in the
       kettle is at 100 °C, what mass of water will turn to steam if it carries on boiling for two
       minutes?

   (2 marks)

Conduction

If you build a house, there are regulations about doing it properly, mainly so that it doesn't fall down, but also so that it <u>keeps the heat in</u>. Easier said than done — there are several ways that heat is 'lost'.

Conduction is most important in solids

Houses lose a lot of heat through their windows even when they're shut. One reason for this is that heat flows from the warm inside face of the window to the cold outside face by <u>conduction</u>.

1) In a <u>solid</u>, the particles are held tightly together. So when one particle <u>vibrates</u>, it <u>bumps into</u> other particles nearby and quickly passes the vibrations on.

2) Particles which vibrate <u>faster</u> than others pass on their <u>extra kinetic energy</u> (that's <u>movement</u> energy) to <u>neighbouring particles</u>. These particles then vibrate faster themselves.

3) This process continues throughout the solid and gradually the extra kinetic energy (or <u>heat</u>) is spread all the way through the solid. This causes a <u>rise in temperature</u> at the <u>other side</u>.

> <u>CONDUCTION OF HEAT</u> is the process where <u>vibrating particles</u> pass on <u>extra kinetic energy</u> to <u>neighbouring particles</u>.

4) <u>Metals</u> are really <u>good conductors of heat</u> — that's why they're used for <u>saucepans</u>.

<u>Non-metals</u> are good for <u>insulating</u> things — e.g. for saucepan <u>handles</u>.

Metals are good conductors because free electrons help "carry" the heat through the metal.

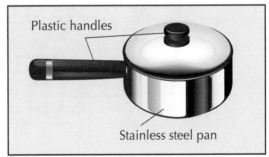

Plastic handles

Stainless steel pan

5) <u>Liquids and gases</u> conduct heat <u>more slowly</u> than solids — the particles aren't held so tightly together. So <u>air</u> is a good insulator.

It's all down to those free electrons

Conduction is like pass the parcel — each particle passes the heat on to its neighbour. They can't chuck heat across the room or anything like that — that's just not playing fair. Each particle vibrates — the more energy each has, the faster it vibrates — and passes energy onto its neighbours.

Convection

Convection occurs in *liquids* and *gases*

1) When you heat up a liquid or gas, the particles move faster, and the fluid (liquid or gas) <u>expands</u>, becoming <u>less dense</u>.

2) The <u>warmer</u>, <u>less dense</u> fluid <u>rises</u> above its <u>colder</u>, <u>denser</u> surroundings, like a hot air balloon does.

3) As the <u>warm</u> fluid <u>rises</u>, cooler fluid takes its place. As this process continues, you actually end up with a <u>circulation</u> of fluid (<u>convection currents</u>). This is how <u>immersion heaters</u> work.

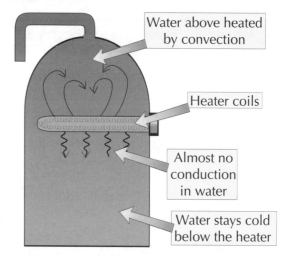

Water above heated by convection

Heater coils

Almost no conduction in water

Water stays cold below the heater

<u>CONVECTION</u> occurs when more energetic particles <u>move</u> from a <u>hotter region</u> to a <u>cooler region</u> — <u>and take their heat energy with them.</u>

4) <u>Radiators</u> in the home rely on convection to make the warm air <u>circulate</u> round the room.

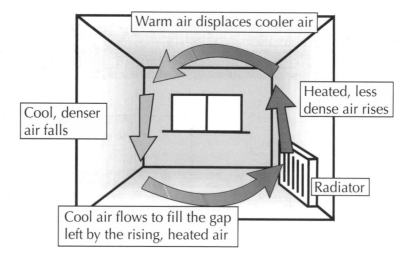

Warm air displaces cooler air

Cool, denser air falls

Heated, less dense air rises

Radiator

Cool air flows to fill the gap left by the rising, heated air

5) Convection <u>can't happen in solids</u> because the <u>particles can't move</u> — they just vibrate on the spot.

6) To <u>reduce convection</u>, you need to <u>stop the fluid moving</u>.

Clothes, blankets and cavity wall foam insulation all work by <u>trapping pockets of air</u>. The air can't move so the heat has to conduct <u>very slowly</u> through the pockets of air, as well as the material in between.

Remember: most 'radiators' don't radiate — they cause convection

If a <u>garden spade</u> is left outside in cold weather, the metal bit will always feel <u>colder</u> than the wooden handle. But it <u>isn't</u> colder — it just <u>conducts heat away</u> from your hand quicker. The opposite is true if the spade is left out in the sunshine — it'll <u>feel</u> hotter because it conducts heat into your hand quicker.

Heat Radiation

The other way heat can be transferred is by <u>radiation</u>.
This is very different from conduction and convection.

*Thermal **radiation** involves **emission** of **electromagnetic waves***

<u>Heat radiation</u> can also be called <u>infrared radiation</u>, and it consists purely of electromagnetic waves of a certain range of frequencies. It's next to visible light in the <u>electromagnetic spectrum</u> (see page 50).

1) <u>All objects</u> continually <u>emit</u> and <u>absorb</u> <u>heat radiation</u>. An object that's <u>hotter</u> than its surroundings <u>emits more radiation</u> than it <u>absorbs</u> (as it <u>cools</u> down). And an object that's <u>cooler</u> than its surroundings <u>absorbs more radiation</u> than it <u>emits</u> (as it <u>warms</u> up).

2) The <u>hotter</u> an object gets, the <u>more</u> heat radiation it <u>emits</u>.

3) You can <u>feel</u> this <u>heat radiation</u> if you stand near something <u>hot</u> like a fire or if you put your hand just above the bonnet of a recently parked car.

(recently parked car)

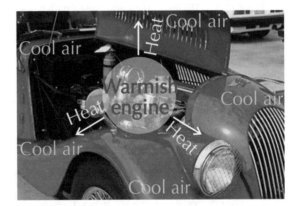

(after an hour or so)

Radiation is how we get heat from the *Sun*

1) <u>Radiation</u> can occur in a <u>vacuum</u>, like space. This is the <u>only way</u> that heat reaches us from the <u>Sun</u>.

2) Heat radiation only passes through substances that are <u>transparent</u> to infrared radiation — e.g. <u>air</u>, <u>glass</u> and <u>water</u>.

3) The <u>amount</u> of radiation emitted or absorbed by an object depends to a large extent on its <u>surface colour and texture</u>. This definitely <u>isn't</u> <u>true</u> for conduction and convection.

If radiation couldn't travel through a vacuum, we'd all be very cold

If you think about sunshine, it's easy to see that heat and light are similar things. They're both part of the electromagnetic spectrum and both travel in waves. The difference between them is their wavelength and frequency — but not their speed, all waves travel at the same speed.

Heat Radiation

The amount of heat **radiated** and **emitted** depends on...

1) Surface area

1) <u>Heat</u> is <u>radiated</u> from the <u>surface</u> of an object.

2) The <u>bigger</u> the <u>surface area</u>, the <u>more waves</u> can be <u>emitted</u> from the surface — so the <u>quicker</u> the <u>transfer of heat</u>.

3) This is why <u>car and motorbike engines</u> often have '<u>fins</u>' — they <u>increase</u> the <u>surface area</u> so heat is radiated away quicker. So the <u>engine cools quicker</u>.

4) It's the same with <u>heating</u> something up — the bigger the surface area exposed to the heat radiation, the <u>quicker it'll heat up</u>.

Cooling fins on engines increase surface area to speed up cooling.

2) Colour and texture

1) <u>Matt black</u> surfaces are very <u>good absorbers and emitters</u> of radiation. Painting a wood-burning stove <u>matt black</u> means it'll <u>radiate</u> as much <u>heat</u> as possible.

2) <u>Light-coloured, smooth</u> objects are very <u>poor absorbers and emitters</u> of radiation, but they effectively <u>reflect</u> heat radiation.

For example, some people put shiny foil behind their radiators to reflect radiation back into the room rather than heat up the walls.

Another good example is <u>survival blankets</u> for people rescued from snowy mountains — their shiny, smooth surface <u>reflects</u> the body heat back inside the blanket, and also <u>minimises</u> heat radiation being <u>emitted</u> by the blanket.

Revise heat radiation — well, at least absorb as much as you can

The most confusing thing about radiation is that those white things on your walls called 'radiators' actually transfer <u>most</u> of their heat by <u>convection</u>, as rising warm air. They do radiate some heat too, of course, but whoever chose the name 'radiator' obviously hadn't swotted up their physics first.

Saving Energy

Insulating *your home* **saves energy** *and* **money**

1) To save energy, you need to insulate your home. It costs money to buy and install the insulation, but it also saves you money, because your heating bills are lower.

2) Eventually, the money you've saved on heating bills will equal the initial cost of installing the insulation — the time this takes is called the payback time.

3) Cheaper methods of insulation are usually less effective — they tend to save you less money per year, but they often have shorter payback times.

4) If you subtract the annual saving from the initial cost repeatedly then eventually the one with the biggest annual saving must always come out as the winner, if you think about it.

5) But you might sell the house (or die) before that happens. If you look at it over, say, a five-year period then a cheap and cheerful hot water tank jacket wins over expensive double glazing.

Loft insulation

Fibreglass 'wool' laid across the loft floor reduces conduction through the ceiling into the roof space.
Initial Cost: £200
Annual Saving: £100
Payback time: 2 years

Hot water tank jacket

Lagging such as fibreglass wool reduces conduction.
Initial Cost: £60
Annual Saving: £15
Payback time: 4 years

Cavity walls & insulation

Two layers of bricks with a gap between them reduce conduction. Insulating foam is squirted into the gap between layers, trapping pockets of air to minimise convection.
Initial Cost: £150
Annual Saving: £100
Payback time: 18 months

The exact costs and savings will depend on the house — but these figures give you a rough idea.

Double glazing

Two layers of glass with an air gap between reduce conduction.
Initial Cost: £2400
Annual Saving: £80
Payback time: 30 years

Draught-proofing

Strips of foam and plastic around doors and windows stop hot air going out — reducing convection.
Initial Cost: £100
Annual Saving: £15
Payback time: 7 years

Thick curtains

Reduce conduction and radiation through the windows.
Initial Cost: £180
Annual Saving: £20
Payback time: 9 years

Thermograms *show where your home is* **leaking heat**

A thermogram is a picture taken with a thermal imaging camera. Objects at different temperatures emit infrared rays of different wavelengths, which the thermogram displays as different colours.

In this thermogram, red shows where most heat is being lost. The houses on the left and right are losing bucket-loads of heat out of their roofs, but the one in the middle must have loft insulation as it's not losing half as much.

TONY MCCONNELL / SCIENCE PHOTO LIBRARY

Warm-Up and Exam Questions

Yes, you've got it — do the warm-up questions first, then when you think you're ready, have a go at the exam questions. If there's anything you can't do, make sure you go back and check on it.

Warm-Up Questions

1) Describe the process of heat transfer by conduction.
2) Explain why heated air rises.
3) Heat radiation can also be called thermal radiation. What is another name for it?
4) Give two ways the nature of a surface could be changed so that the surface emits more heat radiation.
5) Give three methods of insulating a house.

Exam Questions

1   Warm air rises in the roof space of a house.
    What type of heat transfer is this?

   A   radiation

   B   conduction

   C   convection

   D   insulation

   (1 mark)

2   Mandy wants some blinds for her new conservatory so that it doesn't get too hot on very sunny days. Which ones should she choose?

   A   matt, black blinds

   B   matt, white blinds

   C   shiny, black blinds

   D   shiny, white blinds

   (1 mark)

3   The diagram shows the heat losses from Tom's house. Tom estimates that £300 of his annual heating bill is wasted on heat lost from the house.

   (a)   How much money does Tom waste every year in heat lost through the roof?

   (1 mark)

   (b)   Tom decides to insulate the loft. This costs him £350, but reduces the amount he spends on wasted heat to £255 per year.

   Calculate the payback time for fitting loft insulation in Tom's house.

   (2 marks)

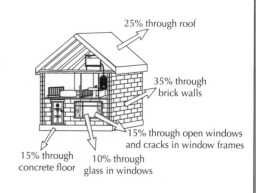

25% through roof

35% through brick walls

15% through open windows and cracks in window frames

10% through glass in windows

15% through concrete floor

Energy

Learn these **nine types** of **energy**

You should know all of these <u>well enough</u> by now to list them <u>from memory</u>, including the examples:

1) <u>ELECTRICAL</u> Energy....................................... — whenever a <u>current</u> flows.
2) <u>LIGHT</u> Energy... — from the <u>Sun</u>, <u>light bulbs</u>, etc.
3) <u>SOUND</u> Energy... — from <u>loudspeakers</u> or anything <u>noisy</u>.
4) <u>KINETIC</u> Energy, or <u>MOVEMENT</u> Energy..... — anything that's <u>moving</u> has it.
5) <u>NUCLEAR</u> Energy.. — released only from <u>nuclear reactions</u>.
6) <u>THERMAL</u> Energy or <u>HEAT</u> Energy.............. — <u>flows</u> from <u>hot objects</u> to colder ones.
7) <u>GRAVITATIONAL POTENTIAL</u> Energy........ — possessed by anything which can <u>fall</u>.
8) <u>ELASTIC POTENTIAL</u> Energy...................... — possessed by <u>springs</u>, <u>elastic</u>, <u>rubber bands</u>, etc.
9) <u>CHEMICAL</u> Energy..................................... — possessed by <u>foods</u>, <u>fuels</u>, <u>batteries</u> etc.

> The <u>last three</u> above are forms of <u>stored energy</u> because the energy is not obviously <u>doing</u> anything, it's kind of <u>waiting to happen</u>, i.e. waiting to be turned into one of the <u>other</u> forms.

There are two types of **"energy conservation"**

Try and get your head round the difference between these two:

1) "<u>ENERGY CONSERVATION</u>" is all about <u>using fewer resources</u> because of the damage they do and because they might <u>run out</u>. That's all <u>environmental stuff</u> — which is important to us, but fairly trivial on a <u>cosmic scale</u>.

2) The "<u>PRINCIPLE OF THE CONSERVATION OF ENERGY</u>", however, is one of the <u>major cornerstones</u> of modern physics. It's an <u>all-pervading principle</u> which governs the workings of the <u>entire physical Universe</u>. If this principle were not so, then life as we know it would simply cease to be.

The **principle of the conservation of energy** can be stated thus:

> <u>Energy</u> can never be <u>created nor destroyed</u>
> — it's only ever <u>converted</u> from one form to another.

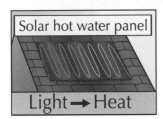

Solar hot water panel

Light → Heat

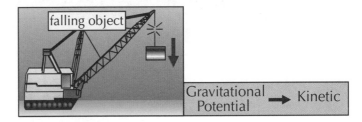

falling object

Gravitational Potential → Kinetic

Another <u>important principle</u> which you need to <u>learn</u> is this one:

> Energy is <u>only useful</u> when it can be <u>converted</u> from one form to another.

Energy Transformation Diagrams

The **thickness** of the **arrow** represents the **amount** of **energy**

The idea of <u>energy transformation (Sankey) diagrams</u> is to make it <u>easy to see</u> at a glance how much of the <u>input energy</u> is being <u>usefully employed</u> compared with how much is being <u>wasted</u>.

The <u>thicker the arrow</u>, the <u>more energy</u> it represents — so you see a big <u>thick arrow going in</u>, then several <u>smaller arrows going off</u> it to show the different energy transformations taking place.

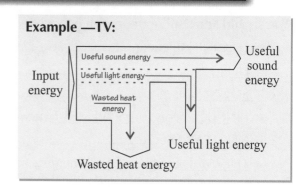

Example — TV:

Sankey diagrams can help you judge **efficiency**

You can have either a little <u>sketch</u> or a properly <u>detailed diagram</u> where the width of each arrow is proportional to the number of joules it represents.

Example — sketch diagram for a simple motor:

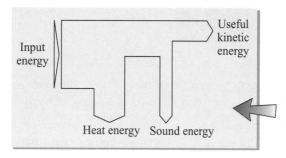

With sketches, they're likely to ask you to <u>compare</u> two different devices and say which is <u>more efficient</u>. You generally want to be looking for the one with the <u>thickest useful energy arrow(s)</u>.

You don't know the actual amounts, but you can see that most of the energy is being <u>wasted</u>, and that it's mostly wasted as <u>heat</u>.

Example — detailed diagram for a simple motor:

In an exam, the most likely question you'll get about detailed Sankey diagrams is filling in one of the numbers or <u>calculating the efficiency</u>. The efficiency is straightforward enough if you can work out the numbers (see next page).

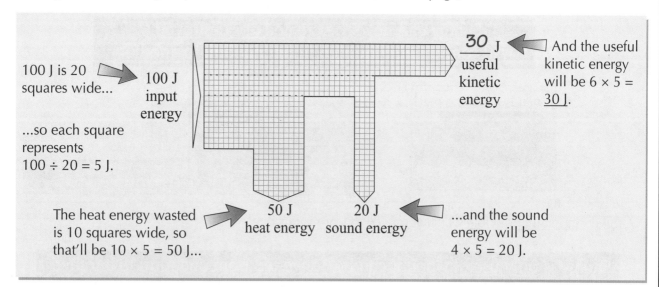

100 J is 20 squares wide...

...so each square represents 100 ÷ 20 = 5 J.

100 J input energy

The heat energy wasted is 10 squares wide, so that'll be 10 × 5 = 50 J...

50 J heat energy

20 J sound energy

...and the sound energy will be 4 × 5 = 20 J.

30 J useful kinetic energy

And the useful kinetic energy will be 6 × 5 = 30 J.

Efficiency

An open fire looks cosy, but a lot of its heat energy goes straight up the chimney, by convection, instead of heating up your living room. All this energy is 'wasted', so open fires aren't very efficient.

Machines *always* **waste** some **energy**

1) <u>Useful machines</u> are only <u>useful</u> because they <u>convert energy</u> from <u>one form</u> to <u>another</u>. Take cars for instance — you put in <u>chemical energy</u> (petrol or diesel) and the engine converts it into <u>kinetic (movement) energy</u>.

2) The <u>total energy output</u> is always the <u>same</u> as the <u>energy input</u>, but only some of the output energy is <u>useful</u>. So for every joule of chemical energy you put into your car you'll only get <u>a fraction of it</u> converted into useful kinetic energy.

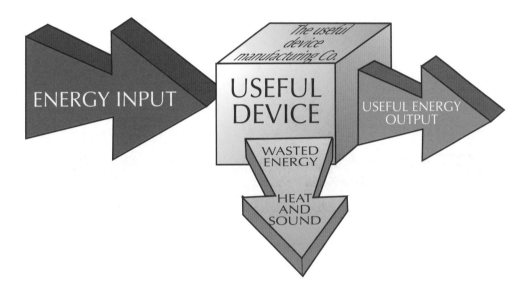

3) This is because some of the <u>input energy</u> is always <u>lost</u> or <u>wasted</u>, often as <u>heat</u>. In the car example, the rest of the chemical energy is converted (mostly) into <u>heat and sound energy</u>. This is wasted energy.

4) The <u>less energy</u> that is <u>wasted</u>, the <u>more efficient</u> the device is said to be.

The **efficiency** of a machine is defined as...

$$\text{Efficiency} = \frac{\text{USEFUL Energy OUTPUT}}{\text{TOTAL Energy OUTPUT}}$$

The input energy is ALWAYS more than the useful energy output

Efficiency is all about what goes in and how it comes out. Remember that conservation of energy thingy — what do you mean no? It's on the last page — anyway that means that however much energy you put in, you'll get that same amount out. The tricky bit is working out how much comes out usefully.

Efficiency

More *efficient* machines *waste less energy*

1) To work out the efficiency of a machine, first find out the Total Energy output.
   This is the same as the energy supplied to the machine — the energy input.

2) Then find how much useful energy the machine delivers — the USEFUL Energy output.
   The question might tell you this directly, or it might tell you how much energy is wasted.

3) Then just divide the smaller number by the bigger one to get a value for efficiency somewhere
   between 0 and 1. Easy. (If your number is bigger than 1, you've done the division upside down.)

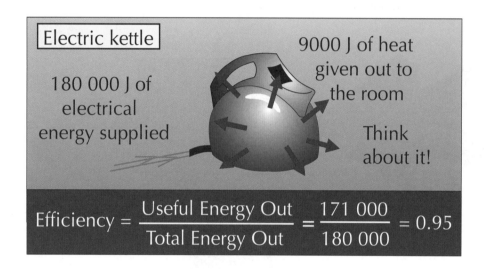

Electric kettle

180 000 J of electrical energy supplied

9000 J of heat given out to the room

Think about it!

$$\text{Efficiency} = \frac{\text{Useful Energy Out}}{\text{Total Energy Out}} = \frac{171\ 000}{180\ 000} = 0.95$$

4) You can convert the efficiency to a percentage, by multiplying it by 100. E.g. 0.6 = 60%.

5) In the exam you might be told the efficiency and asked to work out the total energy output,
   the useful energy output or the energy wasted. So you need to be able to rearrange the formula.

> EXAMPLE: An ordinary light bulb is 5% efficient. If 1000 J of light
> energy is given out, how much energy is wasted?
>
> ANSWER: Total Output$= \dfrac{\text{Useful Output}}{\text{Efficiency}} = \dfrac{1000\ J}{0.05} = 20\ 000\ J,$
> so Energy Wasted $= 20\ 000 - 1000 = \underline{19\ 000\ J}$

Shockingly inefficient, those ordinary light bulbs. Low-energy light bulbs are roughly 4 times more efficient, and last about 8 times as long. They're more expensive though.

Efficiency questions are all more or less the same

Some new appliances (like washing machines and fridges) come with a sticker with a letter from A to H on, to show how energy-efficient they are. A really well-insulated fridge might have an 'A' rating. But if you put it right next to the oven, or never defrost it, it will run much less efficiently than it should.

Warm-Up and Exam Questions

It's no good learning all the facts in the world if you go to pieces and write nonsense in the exam.
So you'd be wise to practise using all your knowledge to answer some questions.

Warm-Up Questions

1) What type of energy is stored in food?
2) State the principle of the conservation of energy.
3) Modern appliances tend to be more energy efficient than older ones. What does this mean?
4) Give an example of a device that uses elastic potential energy.
5) Why is the efficiency of an appliance always less than 100%?

Exam Questions

1   A motor is supplied with 200 J of energy to lift a load.
    The load gains 140 J of potential energy.
    What is the efficiency of the motor?

    A   60 J

    B   70%

    C   0.3

    D   1.43

    (1 mark)

2   Paul is testing the efficiency of four
    different light bulbs. He finds that the
    'Brightlight Ecobulb' is 15% efficient.
    Which of the Sankey diagrams on the
    right could represent this bulb?

    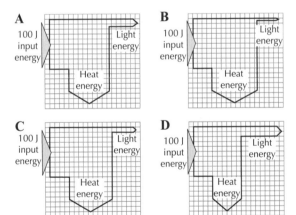

    (1 mark)

3   A hairdryer is supplied with 1200 J of electrical energy every second. The electrical energy
    is converted to 20 J of sound energy and 100 J of kinetic energy every second.

    (a) How much electrical energy does the hairdryer transform into heat energy
        every second?

    (1 mark)

    (b) Suggest how the hairdryer could be made more efficient.

    (1 mark)

Energy Sources

There are various different types of energy resource.
They fit into two broad types: renewable and non-renewable.

Non-renewable energy resources *will* **run out** *one day*

The non-renewables are the three FOSSIL FUELS and NUCLEAR:

1) Coal

2) Oil

3) Natural gas

4) Nuclear fuels (uranium and plutonium)

> a) They will all 'run out' one day.
> b) They all do damage to the environment.
> c) But they provide most of our energy.

There are **environmental problems** *with* **non-renewables**

1) All three fossil fuels (coal, oil and gas) release CO_2. For the same amount of energy produced, coal releases the most CO_2, followed by oil then gas. All this CO_2 adds to the greenhouse effect, and contributes to climate change. We could stop some of it entering the atmosphere — by 'capturing' it and burying it underground, for instance — but the technology is too expensive to be widely used yet.

2) Burning coal and oil releases sulfur dioxide, which causes acid rain.
   This is reduced by taking the sulfur out before it's burned, or cleaning up the emissions.

3) Coal mining makes a mess of the landscape, especially "open-cast mining".

4) Oil spillages cause serious environmental damage. We try to avoid them, but they'll always happen.

5) Nuclear power is clean but the nuclear waste is dangerous and difficult to dispose of (see page 27).

6) But non-renewable fuels are generally concentrated energy resources, reliable, and easy to use.

Learn about the non-renewables — before it's too late

There's lots more info about the various power sources over the next few pages. But the point is that none of them are ideal — they all have pros and cons, and the aim is to choose the least bad option overall. Unless we want to go without heating, light, transport, electricity... and so on. (I don't.)

Energy Sources

Renewable energy sources are often seen as the energy sources of the future — they're more friendly to the environment and won't run out like the non-renewables, but they do have problems of their own...

Renewable *energy resources will* never run out

The renewables are:

1) Geothermal
2) Wind
3) Solar
4) Biomass
5) Waves
6) Tides
7) Hydroelectric

a) These will never run out.
b) Most of them do damage the environment, but in less nasty ways than non-renewables.
c) The trouble is they don't provide much energy and some of them are unreliable because they depend on the weather.

The Sun *is the* ultimate source *of loads of* energy

1) A lot of these energy sources can be traced back to the Sun.

2) Every second for the last few billion years or so, the Sun has been giving out loads of energy — mostly in the form of heat and light.

3) Some of that energy is stored here on Earth as fossil fuels (coal, oil and gas).

> Fossil fuels are the remains of plants and animals that lived long ago.

4) When we use wind power, we're also using the Sun's energy — the Sun heats the air, the hot air rises, cold air whooshes in to take its place (wind), and so on.

Nuclear, geothermal *and* tidal *energy do* not *originate in the Sun*

1) Nuclear power comes from the energy locked up in the nuclei of atoms.

2) Nuclear decay also creates heat inside the Earth for geothermal energy, though this happens much slower than in a nuclear reactor.

3) Tides are caused by the gravitational attraction of the Moon and Sun.

Most energy resources come from the Sun — but not all of them

The Sun really is rather useful — not only does it warm and light our little planet, it also provides lots of energy sources from which we can generate electricity. It's quite obvious that the Sun provides the energy for solar power, but don't forget about wind, waves and biomass — they come from the Sun too.

Nuclear Energy

Well, who'd have thought... there's energy lurking about inside atoms.

Nuclear power uses uranium as fuel

1) A nuclear power station uses uranium to produce heat.

2) Nuclear power stations are expensive to build and maintain, and they take longer to start up than fossil fuel power stations. (Natural gas is the quickest.)

3) Processing the uranium before you use it causes pollution, and there's always a risk of leaks of radioactive material, or even a major catastrophe like at Chernobyl.

4) A big problem with nuclear power is the radioactive waste that you always get. And when they're too old and inefficient, nuclear power stations have to be decommissioned (shut down and made safe) — that's expensive too.

5) But there are many advantages to nuclear power. It doesn't produce any of the greenhouse gases which contribute to global warming. Also, there's still plenty of uranium left in the ground (although it can take a lot of money and energy to make it suitable for use in a reactor).

Nuclear fuel heats water to produce steam

1) The nuclear fuel produces heat.

2) The heat is taken to the steam generator using a liquid coolant.

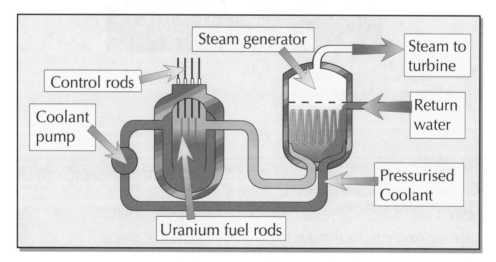

3) The water in the steam generator gets hot, and turns to steam.

4) The steam is used to turn a turbine and produce electricity.

Many people object to nuclear power because of the waste

Nuclear power sounds a bit scary at first, but the method of using it is just the same as for coal and oil and all that boring stuff — you make some heat, which boils water to make steam. I'm even thinking of getting a nuclear powered kettle to save time in the morning when I need that cup of coffee...

Nuclear and Geothermal Energy

Radioactive waste is difficult to dispose of safely

1) Most waste from nuclear power stations and hospitals is 'low-level' (only slightly radioactive). This kind of waste can be disposed of by burying it in secure landfill sites.

2) High-level waste is the really dangerous stuff — a lot of it stays highly radioactive for tens of thousands of years, and so has to be treated very carefully. It's often sealed into glass blocks, which are then sealed in metal canisters. These could then be buried deep underground.

3) However, it's difficult to find suitable places to bury high-level waste. The site has to be geologically stable (e.g. not suffer from earthquakes), since big movements in the rock could disturb the canisters and allow radioactive material to leak out. And even when geologists do find suitable sites, people who live nearby often object.

4) Not all radioactive waste has to be chucked out though — some of it is reprocessed. After reprocessing, you're left with more uranium (for reuse in power stations) and a bit of plutonium (which can be used to make nuclear weapons).

5) But nuclear power stations and reprocessing plants might be a target for terrorists — who could attack the plant, or use stolen material to make a 'dirty bomb'.

Geothermal energy — heat from underground

1) This is only possible in certain places where hot rocks lie quite near to the surface. The source of much of the heat is the slow decay of various radioactive elements including uranium deep inside the Earth.

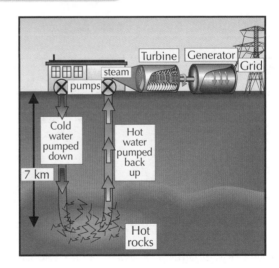

2) Water is pumped in pipes down to hot rocks and returns as steam to drive a generator. This is actually brilliant free energy with no real environmental problems. The main drawback is the cost of drilling down several km.

3) Unfortunately there are very few places where this seems to be an economic option (for now).

It might be expensive — but it'll last forever

Most of the UK's nuclear power stations are quite old, and will be shut down soon. There's a debate going on over whether we should build new ones. Some people say no — if we can't deal safely with the radioactive waste we've got now, we certainly shouldn't make lots more. Others say nuclear power is the only way to meet all our energy needs without causing catastrophic climate change.

Wind and Solar Energy

Wind and solar energy are both renewable — they'll never run out.

Wind farms — lots of little wind turbines

1) Wind power involves putting lots of wind turbines up in <u>exposed places</u> — like on <u>moors</u>, around the <u>coast</u> or <u>out at sea</u>.

2) Wind turbines convert the kinetic energy of moving air into electricity. The <u>wind</u> turns the <u>blades</u>, which turn a <u>generator</u>.

3) Wind turbines are quite cheap to run — they're very <u>tough</u> and reliable, and the wind is <u>free</u>.

4) Even better, wind power doesn't produce any <u>polluting waste</u> and it's <u>renewable</u> — the wind's never going to run out.

5) But there are <u>disadvantages</u>. You normally need a couple of thousand wind turbines to replace one coal-fired power station. Some people think that 'wind farms' spoil the view and the spinning blades cause noise pollution.

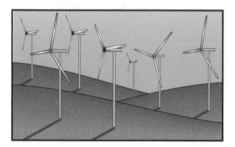

6) Another problem is that sometimes the wind isn't <u>strong enough</u> to generate any power. It's also impossible to increase supply when there's extra demand (e.g. when Coronation Street starts).

7) And although the wind is free, it's <u>expensive</u> to <u>set up</u> a wind farm, especially <u>out at sea</u>.

You can **capture** the Sun's energy using **solar cells**

1) <u>Solar cells</u> (<u>photocells</u>) generate <u>electricity directly</u> from sunlight. They generate <u>direct current</u> (DC) — the same as a <u>battery</u> (not like the <u>mains electricity</u> in your home, which is AC — alternating current).

2) Most solar cells are made of <u>silicon</u> — a <u>semiconductor</u>. When sunlight falls on the cell, the silicon atoms <u>absorb</u> some of the energy, knocking loose some <u>electrons</u>. These electrons then flow round a circuit.

3) The <u>power output</u> of a photocell depends on its <u>surface area</u> (the bigger the cell, the more electricity it produces) and the <u>intensity of the sunlight</u> hitting it (brighter light = more power). Makes sense.

The main problem is the visual impact

Wind turbines — you love them or you hate them. I think they look really good and wouldn't mind seeing a few more. Others can't stand them, while some even have them put on the roof of their house.

Solar Energy

Solar cells — *expensive to install, cheap to run*

1) Solar cells are very <u>expensive initially</u>, but after that the energy is <u>free</u> and <u>running costs</u> are almost <u>nil</u>. And there's <u>no pollution</u> (although they use a fair bit of energy to manufacture in the first place).

2) Solar cells can only <u>generate</u> enough <u>electricity</u> to be useful if they have <u>enough sunlight</u> — which can be a problem at <u>night</u> (and in <u>winter</u> in some places). But the cells can be linked to <u>rechargeable batteries</u> to create a system that can <u>store energy</u> during the day for use at <u>night</u>.

3) Solar cells are often the best way to power <u>calculators</u> or <u>watches</u> that don't use much energy. They're also used in <u>remote places</u> where there's not much choice (e.g. deserts) and in satellites.

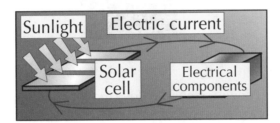

Passive solar heating — *no complex mechanical stuff*

Solar Panels

Solar panels are much less sophisticated than photocells — basically just <u>black water pipes</u> inside a <u>glass</u> box. The <u>glass</u> lets <u>heat</u> and <u>light</u> from the Sun in, which is then <u>absorbed</u> by the black pipes and heats up the water.

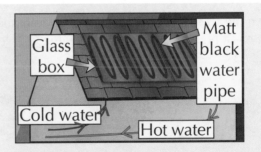

Cooking with Solar Power

If you get a <u>curved mirror</u>, then you can <u>focus</u> the Sun's light and heat. This is what happens in a solar oven.

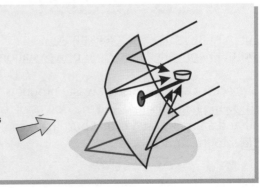

All the radiation that lands on the curved mirror is focused right on your pan.

Unfortunately it's not too reliable here in Britain

And you can reduce the energy needed to <u>heat</u> a building if you build it sensibly in the first place — e.g. face the <u>windows</u> in a suitable direction. That's another example of passive solar heating.

Biomass

Biomass and wave energy sound pretty dull, but they're actually quite exciting possibilities.

Biomass is waste that can be burnt — plant and animal waste

1) Biomass is the general term for organic 'stuff' that can be burnt to produce electricity (e.g. farm waste, animal droppings, landfill rubbish, specially-grown forests...).

Plant material...

...rubbish...

...and waste.

2) The waste material is burnt in power stations to drive turbines and produce electricity. Or sometimes it's fermented to produce other fuels such as 'biogas' (mostly methane) or ethanol.

Biomass is carbon neutral...well, nearly

1) The plants that grew to produce the waste (or to feed the animals that produced the dung) would have absorbed carbon dioxide from the atmosphere as they were growing.

2) When the waste is burnt, this CO_2 is re-released into the atmosphere. So it has a neutral effect on atmospheric CO_2 levels.

Although this only really works if you keep growing plants at the same rate you're burning things, and if you ignore any fossil fuels used in transporting the fuel to the power station, etc.

Biomass schemes are cheap to set up

1) Set-up and fuel costs are generally low, since the fuel is usually waste, and the fuels can often be burnt in converted coal-fired power stations.

2) This process can make use of waste products, which could be great news for our already overflowing landfill sites. But the downside of using unsorted landfill rubbish, rather than just plant and animal waste, is that burning it can release nasty gases like sulfur dioxide and nitrogen oxide into the atmosphere.

Biomass — what a load of old rubbish

Biomass isn't the nicest of things — dead plants, rubbish and poo — mmm, my favourite. It could be a good way of sorting out our energy problems as there's lots of the stuff about already, so don't bin it.

Wave Energy

Wave power — *lots of little* **wave converters**

Don't confuse <u>wave power</u> with <u>tidal power</u> — they're <u>completely different</u>.

1) For wave power, you need lots of small <u>wave converters</u> located <u>around the coast</u>. As waves come in to the shore they provide an <u>up and down motion</u> which can be used to drive a <u>generator</u>.

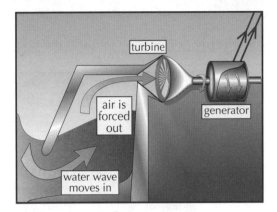

2) There's <u>no pollution</u>. The main problems are <u>spoiling the view</u> and being a <u>hazard to boats</u>.

3) It's <u>fairly unreliable</u>, since waves tend to die out when the <u>wind drops</u>.

4) <u>Initial costs are high</u> but there are <u>no fuel costs</u> and <u>minimal running costs</u>. Wave power is unlikely to provide energy on a <u>large scale</u> but it can be <u>very useful</u> on <u>small islands</u>.

Tidal barrages — *using the* **Sun** *and* **Moon's gravity**

1) <u>Tidal barrages</u> are <u>big dams</u> built across <u>river estuaries</u>, with <u>turbines</u> in them. As the <u>tide comes in</u> it fills up the estuary to a height of <u>several metres</u>. This water can then be allowed out <u>through turbines</u> at a controlled speed. It also drives the turbines on the way in.

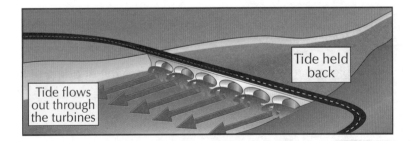

2) There's <u>no pollution</u>. The main problems are <u>preventing free access by boats</u>, <u>spoiling the view</u> and <u>altering the habitat</u> of the wildlife.

3) Tides are <u>pretty reliable</u>, but the <u>height</u> of the tide is <u>variable</u> so lower tides will provide <u>less energy</u> than higher ones.

4) <u>Initial costs are moderately high</u>, but there's <u>no fuel costs</u> and <u>minimal running costs</u>.

Don't get wave power and tidal power confused

I do hope you appreciate the <u>big big differences</u> between <u>tidal power</u> and <u>wave power</u>. They both involve salty sea water, sure — but there the similarities end. Smile and enjoy. And <u>learn</u>.

Hydroelectric Power

Here's another couple of renewables — learn the advantages and disadvantages. They both use water, but hydroelectricity generates electricity, whereas pumped storage only stores it. The clues are in the names — so don't mix them up.

*Hydroelectricity uses **water** to produce **electricity***

1) <u>Hydroelectric power</u> often requires the <u>flooding</u> of a <u>valley</u> by building a <u>big dam</u>.

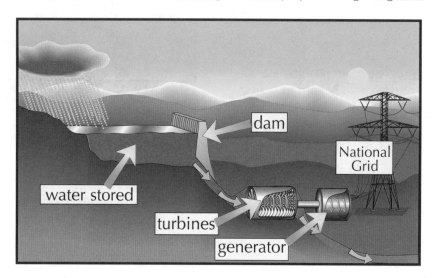

2) <u>Rainwater</u> is caught and allowed out <u>through turbines</u>.

3) The <u>movement</u> of the water turns the <u>turbine</u>, which turns a <u>generator</u> and produces electricity

4) There is <u>no pollution</u> from the running of a hydroelectric scheme.

*Hydroelectricity impacts the **environment***

1) Hydroelectricity has a <u>big impact</u> on the <u>environment</u> due to the flooding of the valley (rotting vegetation releases methane and CO_2) and possible <u>loss of habitat</u> for some species (sometimes the loss of whole villages).

2) The reservoirs can also look very <u>unsightly</u> when they <u>dry up</u>. Location in <u>remote valleys</u> tends to avoid some of these problems.

3) A <u>big advantage</u> is <u>immediate response</u> to increased demand, and there's no problem with <u>reliability</u> except in times of <u>drought</u>.

4) <u>Initial costs are high</u>, but there's <u>no fuel</u> and <u>minimal running costs</u>.

Hydroelectric, unsightly and damaging to the environment, but clean

In Britain only a pretty <u>small percentage</u> of our electricity comes from <u>hydroelectric power</u> at the moment, but in some other parts of the world they rely much more heavily on it. For example, in the last few years, <u>99%</u> of <u>Norway's</u> energy came from hydroelectric power. 99% — that's huge!

Pumped Storage and Power Stations

Pumped storage gives extra supply just when it's needed

1) Most large power stations have <u>huge boilers</u> which have to be kept running <u>all night</u> even though demand is <u>very low</u>. This means there's a <u>surplus</u> of electricity at night.

2) It's surprisingly <u>difficult</u> to find a way of <u>storing</u> this spare energy for <u>later use</u>.

3) <u>Pumped storage</u> is one of the <u>best solutions</u>.

4) In pumped storage, 'spare' <u>night-time electricity</u> is used to pump water up to a <u>higher reservoir</u>.

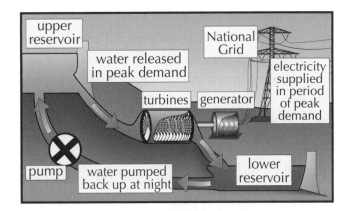

upper reservoir

water released in peak demand

National Grid

electricity supplied in period of peak demand

turbines | generator

pump | water pumped back up at night

lower reservoir

5) This can then be <u>released quickly</u> during periods of <u>peak demand</u> such as at <u>teatime</u> each evening, to supplement the <u>steady delivery</u> from the big power stations.

6) Remember, <u>pumped storage</u> uses the same <u>idea</u> as hydroelectric power but it <u>isn't</u> a way of <u>generating</u> power — but simply a way of <u>storing energy</u> which has <u>already</u> been generated.

Setting up a power station

Because coal and oil are running out fast, many old <u>coal- and oil-fired power stations</u> are being <u>taken out of use</u>. Mostly they're being <u>replaced</u> by <u>gas-fired power stations</u>. But gas is <u>not</u> the <u>only option</u>, as you really ought to know if you've been concentrating at all over the last few pages.

When looking at the options for a <u>new power station</u>, there are <u>several factors</u> to consider:

1) How much it <u>costs</u> to set up and run, <u>how long</u> it takes to <u>build</u>, <u>how much power</u> it can generate, etc.

2) Then there are also the trickier factors like <u>damage to the environment</u> and <u>impact on local communities</u>. And because these are often <u>very contentious</u> issues, getting <u>permission</u> to build certain types of power station can be a <u>long-running</u> process, and hence <u>increase</u> the overall <u>set-up time</u>.

Pumped storage — the clue's in the name, it **stores** energy

Electricity companies have one big problem — customers, they all want their electricity at the same time, so rude. Ok, so it's not the only problem — there's also the fact that coal and oil are running out, nobody wants a nuclear power station in their back yard and wind turbines are ugly. Great.

Warm-Up and Exam Questions

Warm-up questions first, then an exam question or two to practise.
Make the most of this page by working through everything carefully — it's all useful stuff.

Warm-Up Questions

1) Give two ways in which using coal as an energy source causes problems.
2) Describe the major problems with using nuclear fuel to generate electricity.
3) Explain what is meant by biomass.
4) Describe the difference between a pumped storage scheme and a hydroelectric power scheme.
5) Give two reasons why solar cells are not widely used to generate electricity.

Exam Questions

1   Geothermal energy can be described as a renewable energy source.

(a) What is geothermal energy?

(1 mark)

(b) What does 'renewable energy' mean?

(1 mark)

(c) Why is geothermal energy not used much in the U.K.?

(1 mark)

2   Which one of these energy sources does **not** depend on energy radiated by the Sun?

A   Wind

B   Wave

C   Biomass

D   Tidal

(1 mark)

3   The diagram shows a pumped storage plant. Match up the labels **1 – 4** on the diagram with the energy transfers **A, B, C** and **D**.

A   gravitational potential to kinetic

B   kinetic to electrical

C   electrical to kinetic

D   kinetic to gravitational potential

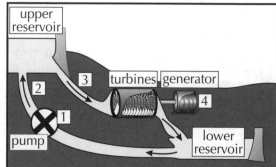

(4 marks)

Exam Questions

4 An old coal-fired power station has an output of 2 MW (2 million watts).
   The electricity generating company plans to replace it with wind turbines which have
   a maximum output of 4000 W each.

   (a) Calculate the minimum number of wind turbines required to replace the
        old power station.

   (1 mark)

   (b) Why might more wind turbines than this be needed in reality?

   (1 mark)

   (c) Suggest why some people might oppose the wind farm development.

   (2 marks)

5 Which is the most reliable method of generating electricity?

   A Tidal power

   B Wind power

   C Solar power

   D Wave power

   (1 mark)

6 The diagram shows a solar heating panel which is used to heat cold water in a house.

   (a) Describe how heat is transferred:

        (i)   from the Sun to the solar heating panel.

        (1 mark)

        (ii) from the hot water in the pipe to the colder
              water in the tank.

        (1 mark)

        (iii) throughout the water in the tank.

        (1 mark)

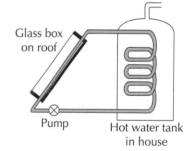

   (b) Explain why the pipes in the heating panel are painted black.

   (1 mark)

7 The inhabitants of a remote island do not have the resources or expertise to build a nuclear
   power plant. They have no access to fossil fuels.

   (a) The islanders have considered using wind, solar and hydroelectric power to generate
        electricity. Suggest two other renewable energy resources they could use.

   (2 marks)

   (b) The islanders decide that both solar and hydroelectric power could reliably generate
        enough electricity for all their needs. Suggest two other factors they should consider
        when deciding which method of electricity generation to use.

   (2 marks)

Revision Summary for Section One

Phew... what a relief, you've made it to the end of yet another nice long section. This one's been fairly straightforward though — after all, about half of it just covered the pros and cons of different renewable energy resources. But don't kid yourself — there are definitely a shedload of facts to remember here and you need to know the lot of them. The best way to check that you know it all is to work your way through these revision questions — and make sure you go back and revise anything you get wrong.

1) Explain the difference between heat and temperature. What units are they each measured in?

2)* A metal rod has a mass of 600 g. It's heated from 18 °C to 28 °C using 5400 J. Calculate the specific heat capacity of the metal.

3) Why does a graph showing the temperature of a substance as it's heated have two flat bits?

4)* The specific latent heat of water (for boiling) is 2.26×10^6 J/kg. How much energy does it take to boil dry a kettle containing 350 g of boiling water?

5) Describe the process that transfers heat energy through a metal rod. What is this process called?

6) Describe how the heat from the element is transferred throughout the water in a kettle. What is this process called?

7) Explain why solar hot water panels have a matt black surface.

8) The two designs of car engine shown are made from the same material. Which engine will transfer heat quicker? Explain why.

Engine A   Engine B

9) Give five ways of reducing the amount of heat lost from a house, and explain how they work.

10)*The following table gives some information about two different energy-saving light bulbs.

a) What is the payback time for light bulb A?
b) Which light bulb is more cost-effective over one year?

	Price of bulb	Annual saving
Light bulb A	£2.50	£1.25
Light bulb B	£3.00	£2.00

11) What does a thermogram show?

12) Name nine types of energy and give an example of each.

13) List the energy transformations that occur in a battery-powered toy car.

14)*The following energy transformation diagram shows how energy is converted in a catapult.

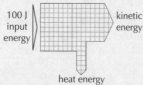
100 J input energy   kinetic energy   heat energy

a) How much energy is converted into kinetic energy?
b) How much energy is wasted?
c) What is the efficiency of the catapult?

15) What is the useful type of energy delivered by a motor? In what form is energy wasted?

16) Write down the formula for calculating efficiency.

17)*What is the efficiency of a motor that converts 100 J of electrical energy into 70 J of kinetic energy?

18) What is meant by a non-renewable energy resource? Name four different non-renewable energy resources.

19) State two advantages and two disadvantages of using fossil fuels to generate electricity.

20) Give two arguments for and two arguments against increasing the use, in the UK, of nuclear power.

21) Give two advantages and one disadvantage of using solar cells to generate electricity.

22) How do solar ovens focus the Sun's rays?

23) Describe how the following renewable resources are used to generate electricity. State one advantage and one disadvantage for each resource.

a) wind   b) biomass   c) geothermal energy
d) waves   e) the tide

*Answers on page 270.

Electric Current

Isn't electricity great — mind you it'll be a pain come exam time if you don't know the basics.

Electric current *is a* flow *of* electrons *round a* circuit

1) <u>CURRENT</u> is the <u>flow of electrons</u> round a circuit. Electrons are <u>negatively charged</u> particles, see p.133.

2) <u>VOLTAGE</u> is the <u>driving force</u> that pushes the current round. Kind of like "<u>electrical pressure</u>".

3) <u>RESISTANCE</u> is anything in the circuit which <u>slows the flow down</u>.

4) There's a <u>BALANCE</u>: the <u>voltage</u> is trying to <u>push</u> the current round the circuit, and the <u>resistance</u> is <u>opposing</u> it — the <u>relative sizes</u> of the voltage and resistance decide <u>how big</u> the current will be:

> If you <u>increase the VOLTAGE</u> — then <u>MORE CURRENT</u> will flow.
>
> If you <u>increase the RESISTANCE</u> — then <u>LESS CURRENT</u> will flow.

It's just like the flow of water *around a set of* pipes

1) The <u>current</u> is simply like the <u>flow of water</u>.

2) <u>Voltage</u> is like the <u>pressure</u> provided by a <u>pump</u> which pushes the stuff round.

3) <u>Resistance</u> is any sort of <u>constriction</u> in the flow, which is what the pressure has to <u>work against</u>.

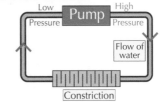

4) If you <u>turn up the pump</u> and provide more <u>pressure</u> (or "<u>voltage</u>"), the flow will <u>increase</u>.

5) If you put in more <u>constrictions</u> ("<u>resistance</u>"), the flow (current) will <u>decrease</u>.

Electrons *flow the* opposite way *to* conventional current

We <u>normally</u> say current in a circuit flows from <u>positive to negative</u>. Alas, electrons were discovered long after that was decided and they turned out to be <u>negatively charged</u>. This means they <u>actually flow</u> from −ve to +ve, <u>opposite</u> to the flow of "<u>conventional current</u>".

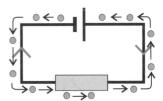

AC *keeps* changing direction *but* DC *does* not

1) The <u>mains electricity</u> supply in your home is <u>alternating current</u> — <u>AC</u>. It keeps <u>reversing its direction</u> back and forth.

2) A cathode ray oscilloscope (CRO) can show current as a trace on a graph. The CRO trace for AC would be a <u>wave</u>. The voltage rises from zero to a <u>peak</u> positive value, then drops down to a 'peak' negative value, then back up to zero again, and so on. The <u>frequency</u> of the supply is <u>how many</u> of these <u>waves</u> you get <u>per second</u>.

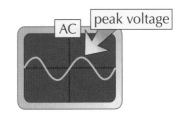

3) <u>Direct current (DC)</u> is <u>different</u>. It <u>always flows</u> in the <u>same direction</u>.

4) The <u>CRO</u> trace is a <u>horizontal line</u>. The <u>voltage doesn't vary</u> — so the <u>current</u> has a <u>constant</u> value too.

5) You get <u>DC current</u> from <u>batteries</u> and <u>solar cells</u> (see page 142).

Current, Voltage and Resistance

Resistance, current and voltage are all closely linked. And if you don't believe me, you can easily check.

Investigating how current varies with voltage

1) This circuit can be used to investigate how current varies with voltage for any component.

2) The ammeter is always connected in series with the component you're testing, to measure current flowing through it.

3) The voltmeter is always connected in parallel with the component to measure the voltage across it.

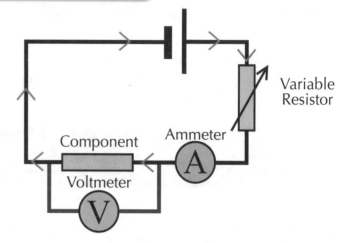

4) The supply voltage from the cell doesn't change. You use a variable resistor in the circuit, which you adjust to pick different values for the current, and for each value measure the voltage across the component.

You can use a V-I graph to look at resistance

Data from a circuit like the one above can be plotted it on a V-I (voltage-current) graph. This shows what happens to the current when you vary the voltage — as a bonus, the gradient shows the resistance.

Fixed-value resistor

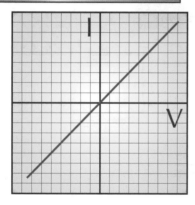

1) The resistance, R, of the component is constant (at a constant temperature).
2) If you plot voltage against current, you get a straight line — so you can see that the current is proportional to the voltage.

Filament Lamp

1) With a filament lamp, the resistance isn't constant — it increases as the current increases. This is because the bigger the current through the filament lamp, the hotter it gets. And as its temperature increases, its resistance increases.
2) The graph's a curve — the current is not proportional to the voltage.

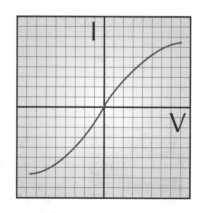

Current, Voltage and Resistance

*You need to know the **resistance equation***

1) The <u>current</u> that flows through a component always depends on the <u>resistance</u> of the component and the <u>voltage</u> across it. There's a very simple equation for it:

$$\text{Voltage} = \text{Current} \times \text{Resistance}$$
(volts, V)   (amps, A)   (ohms, Ω)

or this much shorter version...

$$V = I \times R$$

V is Voltage, and R is Resistance, but, oddly, I is for Current.

2) The equation above has <u>voltage</u> as its subject. But you might need to find <u>resistance</u> or <u>current</u> — which means rearranging the equation. You can do this with a <u>formula triangle</u>.

Cover up the thing you're trying to find, then what you can still see is the formula you need to use.

*Conditions affect the resistance of **LDRs** and **thermistors***

1) A <u>light-dependent resistor</u> (LDR) has a resistance which depends on the <u>light level</u>.

2) The resistance is <u>highest</u> in total <u>darkness</u>. As the light gets <u>brighter</u>, the <u>resistance falls</u>. This makes it a useful device for various <u>electronic sensors</u>, e.g. <u>automatic night lights</u> and <u>burglar detectors</u>.

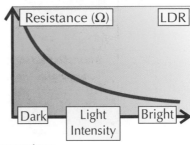

3) Thermistors have varying resistance which depends on the <u>temperature</u>.

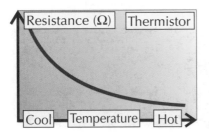

4) Most thermistors have a <u>high</u> resistance in <u>cold</u> conditions. As it gets <u>warmer</u>, the <u>resistance falls</u>. Thermistors are very useful as <u>temperature sensors</u> — e.g. in <u>car-engine</u> temperature gauges and <u>central-heating thermostats</u>.

In the end, you'll have to learn this — resistance is futile...
<u>V = I × R</u> is without doubt the most important equation in electronics. LEARN IT LEARN IT LEARN IT.

The Dynamo Effect

Generators use a pretty cool piece of physics to make electricity from the movement of a turbine. It's called electromagnetic (EM) induction — which basically means making electricity using a magnet.

> ELECTROMAGNETIC INDUCTION: The creation of a VOLTAGE (and maybe current) in a wire which is experiencing a CHANGE IN MAGNETIC FIELD.

The dynamo effect — move the wire or the magnet

1) Using electromagnetic induction to transform kinetic energy (energy of moving things) into electrical energy is called the dynamo effect. (In a power station, this kinetic energy is provided by the turbine.)

2) There are two different situations where you get EM induction:
   a) An electrical conductor (a coil of wire is often used) moves through a magnetic field.

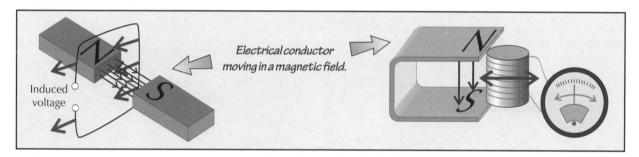

Induced voltage

Electrical conductor moving in a magnetic field.

   b) The magnetic field through an electrical conductor changes (gets bigger or smaller or reverses).

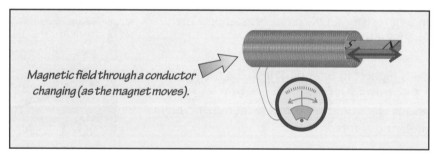

Magnetic field through a conductor changing (as the magnet moves).

3) If the direction of movement is reversed, then the voltage/current will be reversed too.

> To get a bigger voltage, you can increase...
> 1) The STRENGTH of the MAGNET
> 2) The number of TURNS on the COIL
> 3) The SPEED of movement

Learn the three ways of getting a bigger voltage

Who'd have thought you could make electricity from a magnet and a bit of wire — try it yourself if you don't believe it. While you're there try changing the magnet and the number of turns and things to see how the voltage changes. Once you get the hang of it, it'll stick in your head for the exam no trouble.

The Dynamo Effect

Generators *move a coil in a* **magnetic field**

1) Generators usually <u>rotate a coil</u> in a <u>magnetic field</u>.

2) Every half a turn, the current in the coil <u>swaps direction</u>.

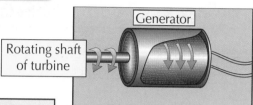

Think about one part of the coil... sometimes it's heading for the magnet's north pole, sometimes for the south — it changes every half a turn. This is what makes the current change direction.

3) This means that generators produce an <u>alternating (AC) current</u>. If you looked at the current (or voltage) on a display, you'd see something like this...

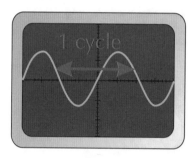
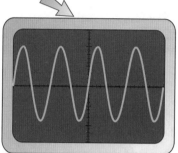

1 cycle

Turning the coil <u>faster</u> produces not only <u>more</u> peaks, but a <u>higher voltage</u> too.

4) The <u>frequency</u> of AC electrical supplies is the number of 'cycles' per second, and is measured in <u>hertz</u> (Hz). In the UK, electricity is supplied at 50 Hz (which means the coil in the generator at the power station is rotating 50 times every second).

5) Remember, this is completely different from the DC electricity supplied by batteries and solar cells. If you plotted that on a graph, you'd see something more like this...

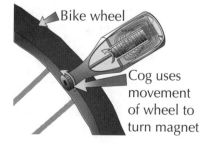
Bike wheel

Cog uses movement of wheel to turn magnet

6) <u>Dynamos</u> on bikes work slightly differently — they usually rotate the <u>magnet</u> near the coil. But the principle is <u>exactly the same</u> — they're still using EM induction.

Generators rotate a coil — dynamos rotate the magnet

EM induction sounds pretty hard, but it boils down to this — if a <u>magnetic field changes</u> (moves, grows, shrinks... whatever) somewhere near a <u>conductor</u>, you get <u>electricity</u>. It's a weird old thing, but important — this is how all our mains electricity is generated. We'd be in the dark without it.

Power Stations and the National Grid

Most of the electricity you use arrives via the national grid.

The *national grid* connects *power stations* to *consumers*

1) The <u>national grid</u> is the <u>network</u> of pylons and cables which covers <u>the whole country</u>.

2) It takes electricity from <u>power stations</u> to just where it's needed in <u>homes</u> and <u>industry</u>.

3) It enables power to be <u>generated</u> in a power station anywhere on the grid, and then <u>supplied</u> anywhere else on the grid.

All *power stations* are pretty much the *same...*

1) The aim of a <u>power station</u> is to <u>convert</u> one kind of energy (e.g. the energy stored in fossil fuels, or nuclear energy contained in the centre of atoms) into <u>electricity</u>.

2) Usually this is done in <u>three stages</u>...

① The first stage is to use the <u>fuel</u> (e.g. gas or nuclear fuel) to generate <u>steam</u> — this is the job of the <u>boiler</u>.

② The moving steam drives the blades of a <u>turbine</u>...

③ ...and this rotating movement from the turbine is converted to <u>electricity</u> by the <u>generator</u> (using <u>electromagnetic induction</u> — see the previous page).

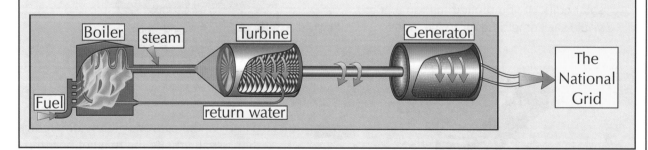

3) Most power stations are terribly <u>inefficient</u> — usually more than half the energy produced is <u>wasted</u> as <u>heat and noise</u> (though the efficiency of the power station depends a lot on the <u>power source</u>).

Most power stations are very very inefficient

Power stations might be big, but they're not all that clever — they boil water to make steam to turn a turbine. They also waste a lot of energy, which isn't great. So, invent a 100% efficient power station and you'll be rich and famous — I guess it's quite tricky though, or someone would've already done it.

Power Stations and the National Grid

The electricity generated by power stations is distributed around the country in the national grid. While it is being transmitted it has a higher voltage than in the power station and in your house.

Electricity is **transformed** to **high voltage** before **distribution**

1) To transmit a lot of electrical power, you either need a <u>high voltage</u> or a <u>high current</u> (see page 182 for more info about why). But... a higher current means your cables get hot, which is very inefficient (all that heat just goes to waste).

2) It's much cheaper to <u>increase</u> the <u>voltage</u>. So before the electricity is sent round the country, the voltage is <u>transformed</u> (using a <u>transformer</u>) to <u>400 000 V</u>. This keeps the current very <u>low</u>, meaning less wasted energy.

3) To increase the voltage, you need a <u>step-up transformer</u>.

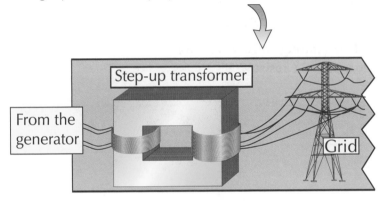

4) Even though you need big <u>pylons</u> with <u>huge</u> insulators (as well as the transformers themselves), using a high voltage is the <u>cheapest</u> way to transmit electricity.

5) To bring the voltage down to <u>safe usable levels</u> for homes, there are local <u>step-down transformers</u> scattered round towns — for example, look for a little fenced-off shed with signs all over it saying "Keep Out" and "Danger of Death".

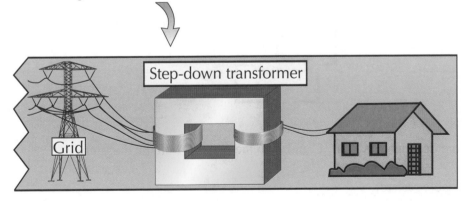

6) This is the main reason why mains electricity is AC — transformers <u>only work</u> on <u>AC</u>.

Fuel, boiler, steam, turbine, generator, transformer, grid, toaster

If you had your own solar panel or wind generator, you could sell back any surplus electricity to the national grid. So if you don't use much electricity, but you generate a lot of it, you can actually make money instead of spending it. Nice trick if you can do it. Shame solar panels cost a fortune...

Warm-Up and Exam Questions

This stuff isn't everyone's cup of tea. But once you get the knack of it, through lots of practice, you'll find the questions aren't too bad. Which is nice.

Warm-Up Questions

1) What is electrical current?
2) Explain the difference between AC and DC.
3) What happens to the resistance of a thermistor as it gets cooler?
4) What is electromagnetic induction?
5) What is the National Grid?
6) Write down the equation that relates resistance, current and voltage.

Exam Questions

1   A current of 2.5 A flows through a 10 W resistor.
     What is the voltage across the resistor?

     A   25 V

     B   10 V

     C   4 V

     D   2.5 V

(1 mark)

2   The circuit shown can be used to find the resistance of a component.

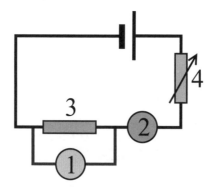

     Match the descriptions **A**, **B**, **C** and **D** to the parts **1-4** on the diagram.

     A   Variable resistor

     B   Voltmeter

     C   Ammeter

     D   Component to be tested

(4 marks)

Exam Questions

3   Electricity is transmitted over long distances in the National Grid at

   A   low voltage and high current.

   B   low voltage and low current.

   C   high voltage and high current.

   D   high voltage and low current.

(1 mark)

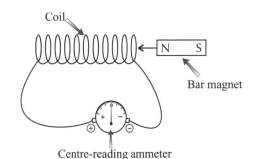

Coil

Bar magnet

4   The diagram shows a coil of wire connected to an ammeter. Tim moves a bar magnet into the coil as shown. The pointer on the ammeter moves to the left.

Centre-reading ammeter

  (a)   Explain why the pointer moves.

(1 mark)

  (b)   What could Tim do to get the ammeter's pointer to move to the right?

(1 mark)

  (c)   How could Tim get a larger reading on the ammeter?

(1 mark)

  (d)   What reading will the ammeter show if Tim holds the magnet still inside the coil?

(1 mark)

5   The diagram shows how heat energy from burning coal is used to generate electricity.

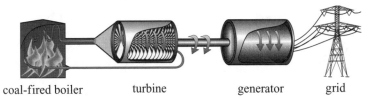

coal-fired boiler   turbine   generator   grid

  (a)   In what form is the energy in coal stored?

(1 mark)

  (b)   Coal has an energy content of approximately 25 000 kJ/kg.

    (i)   How much heat energy is released in burning 1000 kg of coal?

(1 mark)

    (ii)   Calculate the percentage efficiency of the power station if 10 000 MJ of electricity is generated using 1000 kg of coal.

(1 mark)

Electrical Power

This page is about the <u>power</u> of electrical appliances.

Electrical power is the rate of transfer of electrical energy

1) Electrical appliances are useful because they take in <u>electrical energy</u> and <u>convert it</u> into <u>other forms of energy</u>, e.g. a light bulb turns <u>electrical</u> energy into <u>light</u> and <u>heat</u> energy.

2) Converting energy from one form to another is called <u>transfer of energy</u>.

3) The electrical <u>power</u> of an appliance tells you how <u>quickly</u> it transfers electrical energy. The <u>units</u> of power are watts (W) or kilowatts (kW). 1 kilowatt = 1000 watts.

> ## <u>ELECTRICAL POWER</u> is the <u>Rate of Transfer</u> of <u>Electrical Energy</u>.

4) The <u>higher</u> the power of your appliance, the <u>more energy</u> is transferred every second. So a 3 kW kettle boils water <u>faster</u> than a 2 kW kettle, and a 100 W light bulb is <u>brighter</u> than a 60 W bulb.

Power = current × voltage

There's a nice easy equation for the <u>power</u> of an appliance...

> ## POWER = CURRENT × VOLTAGE
> (watts, W)   (amps, A)   (volts, V)

...or this much shorter version. $P = I \times V$

As usual, you need to practise <u>rearranging</u> the equation too.

$$\frac{P}{I \times V}$$

| EXAMPLE: | Anna's hairdrier has a power rating of 1.1 kW. She plugs the hairdrier into the 230 V mains supply. What is the current through the hairdrier? |

You'll need to change the <u>units</u> — in this formula, power has to be in <u>watts</u>, not kilowatts.

| ANSWER: | You're trying to find <u>current</u>, so you need to rearrange the equation. Using the formula triangle, I = P ÷ V = 1100 ÷ 230 = <u>4.8 A</u>. |

Electrical Power

All the electricity you use has to be paid for. Your electricity meter counts how many units of electricity you use and the electricity company multiplies this by the cost of each one to work out your bill.

Kilowatt-hours (kWh) are "UNITS" of energy

Your electricity meter records how much <u>energy</u> you use in units of <u>kilowatt-hours</u>, or <u>kWh</u>.

> A <u>KILOWATT-HOUR</u> is the amount of electrical energy converted by a <u>1 kW appliance</u> left on for <u>1 HOUR</u>.

Appliances with **high power ratings** cost more to run

The <u>higher</u> the <u>power rating</u> of an appliance, and the <u>longer</u> you leave it on, the more energy it consumes, and the more it costs. Learn (and practise rearranging) this equation too...

> UNITS OF ENERGY = POWER × TIME
> (in kWh)   (in kW)   (in hours)

Write out a formula triangle if it helps.

And this one (but this one's easy):

> COST = NUMBER OF UNITS × PRICE PER UNIT

> <u>EXAMPLE:</u>   Find the cost of leaving a 60 W light bulb on for 30 minutes if one kWh costs 10p.

> <u>ANSWER:</u>   Energy (in kWh) = Power (in kW) × Time (in hours)
> = 0.06 kW × ½ hr = <u>0.03 kWh</u>
> Cost = number of units × price per unit = 0.03 × 10p = <u>0.3p</u>

Watt is the unit of power?

Get a bit of <u>practice</u> with the equations in those lovely burgundy boxes, and try these questions:
1) A kettle draws a current of 12 A from the 230 V mains supply. Calculate its power rating.
2) With 0.5 kWh of energy, for how long could you run the kettle?   Answers on p.270.

Warm-Up and Exam Questions

I know that you'll be champing at the bit to get into the exam questions,
but these warm-up questions are invaluable for getting the basics straight first.

Warm-Up Questions

1) What is electrical power?
2) What unit is used by electricity suppliers to charge customers for electricity usage?
3) Write down the formula linking voltage, current and power.
4) Calculate the energy used by a 3.1 kW kettle if it's on for 1½ minutes.

Exam Questions

1   A 1200 W toaster is used with the 230 V mains electricity supply.
    Calculate the current it draws.

    A   0.19 A

    B   1 A

    C   5.22 A

    D   13 A

(1 mark)

2   James's washing machine has a power rating of 800 W. His electricity company charges
    14p per kWh. How much will the electricity for a 45-minute washing cycle cost?

    A   84p

    B   8.4p

    C   50.4p

    D   5.4p

(1 mark)

3   Sarah moves into her new flat and wants to estimate her electricity bill.
    She writes down the power of all her electrical appliances and estimates the time
    they're on for each day. She finds that she uses an average of 8.75 kWh per day.

    (a) How much energy, in joules, does she use per day?

(1 mark)

    (b) If electricity costs 14.4p per kWh, what will her daily cost be?

(1 mark)

    (c) The electricity company sends Sarah a bill every three months.
        Calculate her likely bill for 90 days' electricity supply.

(1 mark)

    (d) A new electricity supplier charges 12p per kWh for electricity.
        How much would Sarah save per day if she switched to this supplier?

(1 mark)

Waves — The Basics

Waves transfer <u>energy</u> from one place to another without transferring any <u>matter</u> (stuff).

Waves have amplitude, wavelength and frequency

1) The <u>amplitude</u> is the displacement from the <u>rest position</u> to the <u>crest</u> (NOT from a trough to a crest).

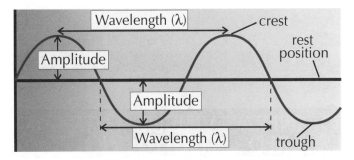

2) The <u>wavelength</u> is the length of a <u>full cycle</u> of the wave, e.g. from <u>crest to crest</u>.

3) <u>Frequency</u> is the <u>number of complete waves</u> passing a certain point <u>per second</u>. Frequency is measured in hertz (Hz). 1 Hz is <u>1 wave per second</u>.

Transverse waves have sideways vibrations

<u>Most waves</u> are <u>transverse</u>:

1) <u>Light</u> and <u>all other EM waves</u>.

2) <u>Ripples</u> on water.

3) <u>Waves</u> on <u>strings</u>.

4) A <u>slinky spring</u> wiggled up and down.

In <u>TRANSVERSE</u> waves the vibrations are at <u>90°</u> to the <u>DIRECTION OF TRAVEL</u> of the wave.

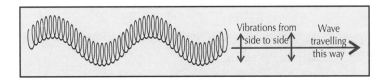

Longitudinal waves have vibrations along the same line

Examples of <u>longitudinal waves</u> are:

1) <u>Sound waves</u> and <u>ultrasound</u>.

2) <u>Shock waves</u>, e.g. seismic waves (see p.67).

3) A <u>slinky spring</u> when you <u>push</u> the end.

In <u>LONGITUDINAL</u> waves the vibrations are along the <u>SAME DIRECTION</u> as the wave is travelling.

Oscilloscopes show longitudinal waves as transverse — just so you can see what's going on.

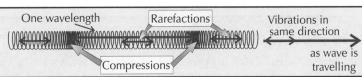

Electromagnetic Waves

These types of wave come up loads in physics so you'd better get them sorted in your head.

There are **seven types** of electromagnetic (EM) waves

Electromagnetic (EM) radiation is all around you. There are <u>seven</u> basic types of electromagnetic waves:

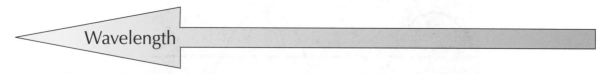

Wavelength ←						
RADIO WAVES	MICRO WAVES	INFRA RED	VISIBLE LIGHT	ULTRA VIOLET	X-RAYS	GAMMA RAYS
$1m\text{-}10^{4}m$	$10^{-2}m$ (3cm)	$10^{-5}m$ (0.01mm)	$10^{-7}m$	$10^{-8}m$	$10^{-10}m$	$10^{-12}m$

Frequency →

Electromagnetic waves travel at the **same speed**

1) All forms of electromagnetic radiation travel at the <u>same speed through a vacuum</u>.

2) Waves with a <u>shorter wavelength</u> have a <u>higher frequency</u> (see next page for why).

3) As a rule the EM waves at <u>each end</u> of the spectrum tend to be able to <u>pass through material</u>, while those <u>nearer the middle</u> are <u>absorbed</u>.

> When EM radiation is <u>absorbed</u>, it can cause:
>
> i) <u>heating</u>,
>
> ii) a tiny <u>AC current</u> with the same frequency as the radiation.

4) Also, the ones with <u>higher frequency</u> (shorter wavelength), like X-rays, tend to be <u>more dangerous</u> to living cells. That's because they have more energy. See page 57 for more information.

5) About half the EM radiation we receive from the <u>Sun</u> is <u>visible light</u>. Most of the rest is <u>infrared</u> (heat), with some <u>UV</u> thrown in. UV is what gives us a suntan (see page 58).

Remember: all waves carry energy without transferring matter

Waves carry energy, but can also carry <u>information</u> — e.g. EM waves carry TV signals, sound waves carry speech, and water waves carry... um... boats. Anyway, get learning about the properties of electromagnetic waves. Quite a straightforward page, so make the most of it.

Wave Speed and Interference

This stuff is true for __all__ waves — not just EM ones...

Wave speed = frequency × wavelength

You need to learn this equation (it's not given in the exam) and <u>practise using it</u>.

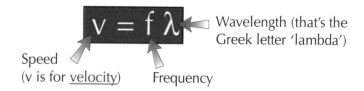

$$\text{Speed} = \text{Frequency} \times \text{Wavelength}$$
$$\text{(m/s)} \qquad \text{(Hz)} \qquad \text{(m)}$$

Or you can use the shortened version:

$$v = f\lambda$$

Wavelength (that's the Greek letter 'lambda')

Speed (v is for <u>velocity</u>)   Frequency

> EXAMPLE: A radio wave has a frequency of 92.2×10^6 Hz.
> Find its wavelength.
> (The speed of all EM waves is 3×10^8 m/s.)

> ANSWER: You're trying to find λ using f and v, so you've got to rearrange the equation.
> So $\lambda = v \div f = 3 \times 10^8 \div 9.22 \times 10^7 = \underline{3.25 \text{ m}}$.

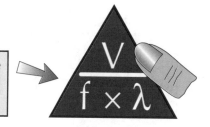

$$\dfrac{V}{f \times \lambda}$$

All EM waves travel at the same speed in a vacuum — so <u>waves with a high frequency must have a short wavelength</u>.

Waves can interfere with each other

1) When two or more waves of a <u>similar frequency</u> come into contact, they can create one combined signal with a new <u>amplitude</u>.

2) This is called <u>interference</u>. You get it when two radio stations transmit on similar frequencies.

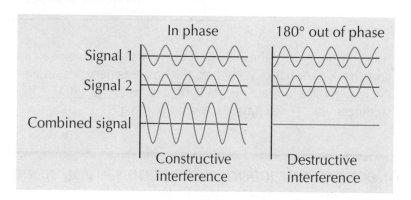

In phase | 180° out of phase

Signal 1
Signal 2
Combined signal

Constructive interference | Destructive interference

Diffraction

Waves travel along by themselves quite happily — but what happens when they meet an obstacle...

All waves can be reflected, refracted and diffracted

When waves arrive at an obstacle (or meet a new material), their direction of travel can be changed.

1) The waves might be <u>reflected</u> — so the waves 'rebound off' the material.

2) They could be <u>refracted</u> — which means they go through the new material but <u>change direction</u> (page 53).

3) Or they could be <u>diffracted</u> — this means the waves 'bend round' obstacles, causing the waves to spread out. This allows waves to 'travel round corners'.

Here's how diffraction works:

- All waves (<u>diffract</u>) <u>spread out</u> at the edges when they pass through a <u>gap</u> or <u>past an object</u>.

- The amount of diffraction depends on the size of the gap relative to the wavelength of the wave. The <u>narrower the gap</u>, or the <u>longer the wavelength</u>, the <u>more</u> the wave spreads out.

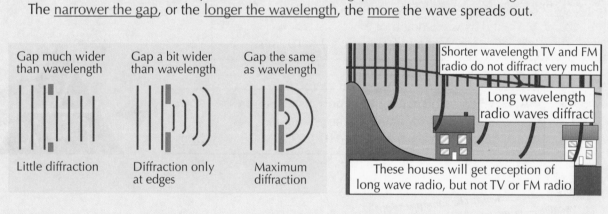

Gap much wider than wavelength — Little diffraction

Gap a bit wider than wavelength — Diffraction only at edges

Gap the same as wavelength — Maximum diffraction

Shorter wavelength TV and FM radio do not diffract very much

Long wavelength radio waves diffract

These houses will get reception of long wave radio, but not TV or FM radio

4) As well as changing a wave's direction of travel, an obstacle can have an effect on the wave itself. The obstacle might:

> i) <u>absorb</u> the wave,
>
> ii) <u>transmit</u> it (let it pass through),
>
> iii) or <u>reflect</u> it.

Any <u>combination</u> of these three is also possible (so you can look at stuff in shop windows <u>and</u> check your hair). What actually happens depends on the <u>wavelength</u> of the radiation, the <u>material</u> the obstacle is made of, and what the <u>surface</u> of this material is like (its colour, shininess, etc. — see page 16 for more info).

Reflection, refraction, diffraction — you need to know them all

In 1588, <u>beacons</u> were used on the south coast of England to <u>relay</u> the information that the Spanish Armada was approaching. As we know, <u>light</u> travels as <u>electromagnetic waves</u>, so this is an early example of transferring information using electromagnetic radiation — or <u>wireless communication</u>.

Refraction

All waves can be <u>refracted</u> — it's a fancy way of saying '<u>made to change direction</u>'.

*Waves can be **refracted***

1) Waves travel at <u>different speeds</u> in substances which have <u>different densities</u>. EM waves travel more <u>slowly</u> in <u>denser</u> media (usually). Sound waves travel faster in <u>denser</u> substances.

2) Here's an example of an EM wave crossing a boundary between two substances and <u>changing speed</u>.

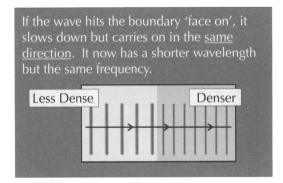

If the wave hits the boundary 'face on', it slows down but carries on in the <u>same direction</u>. It now has a shorter wavelength but the same frequency.

Less Dense | Denser

3) When light shines on a glass <u>window pane</u>, some of the light is reflected, but a lot of it passes through the glass and gets <u>refracted</u> as it does so.

4) As the light passes from the air into the glass (a <u>denser</u> medium), it <u>slows down</u>. This causes the light ray to bend <u>towards</u> the normal.

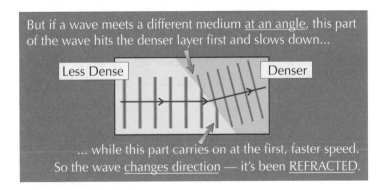

But if a wave meets a different medium <u>at an angle</u>, this part of the wave hits the denser layer first and slows down...

Less Dense | Denser

... while this part carries on at the first, faster speed. So the wave <u>changes direction</u> — it's been <u>REFRACTED</u>.

5) When the light reaches the 'glass to air' boundary on the other side of the window, it <u>speeds up</u> and bends <u>away</u> from the normal. (Some of the light is also <u>reflected</u> at each of the boundaries.)

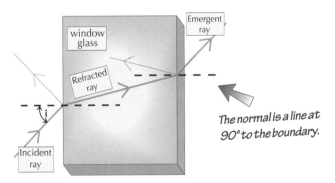

Emergent ray

window glass

Refracted ray

The normal is a line at 90° to the boundary.

Incident ray

Make sure you can describe refraction and all the diagrams

Refraction sounds a bit like reflection — but they're not the same thing, so try not to mix them up. Refraction is all about waves travelling through different things, whereas reflection is where the waves bounce off a surface and come right back at you. See, not the same thing at all.

Refraction

Total internal reflection happens *above* the *critical angle*

1) Total internal reflection can only happen when a wave travels through a dense substance like glass or water or perspex towards a less dense substance like air.

2) It all depends on whether the angle of incidence (i.e. the angle it hits at) is bigger than the critical angle...

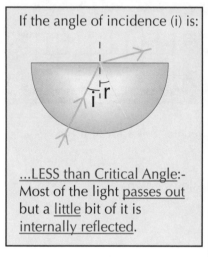

If the angle of incidence (i) is:

...LESS than Critical Angle:-
Most of the light passes out but a little bit of it is internally reflected.

Angle of reflection, r, equals the angle of incidence, i.

The angle of incidence (i) and the angle of reflection (r) are always measured from the normal (a line at right angles to the surface).

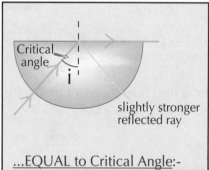

Critical angle

slightly stronger reflected ray

...EQUAL to Critical Angle:-
The emerging ray comes out along the surface. There's quite a bit of internal reflection.

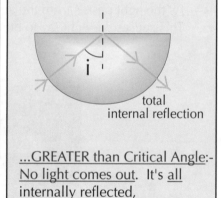

total internal reflection

...GREATER than Critical Angle:-
No light comes out. It's all internally reflected, i.e. total internal reflection.

3) Different materials have different critical angles. The critical angle for glass is about 42°.

4) Optical fibres work by bouncing visible or infrared light waves off the sides of a thin inner core of glass or plastic using total internal reflection. The wave enters one end of the fibre and is reflected repeatedly until it emerges at the other end.

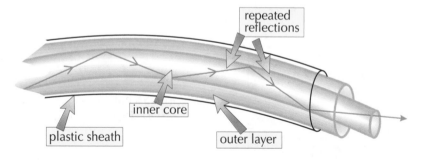

repeated reflections

inner core

plastic sheath

outer layer

5) Optical fibres can be bent, but not sharply, or the angle of incidence might fall below the critical angle.

Total internal reflection can be amazingly useful
Optical fibres are a good way to send data over long distances — the EM waves travel fast, and they can't be tapped into or suffer interference (unlike a signal that's broadcast from a transmitter, like radio).

Warm-Up and Exam Questions

Warm-Up Questions

1) What happens when a wave is refracted, and why?
2) What is meant by the frequency of a wave?
3) Explain how waves 'interfere' with each other, and how this affects the signals they're carrying.
4) Waves spread out when they travel through gaps. What is this effect called?
5) In what circumstances does total internal reflection of light occur?
6) What is an optical fibre? Why don't they work if they're bent sharply?

Exam Questions

1   Which of the following types of electromagnetic wave has the highest frequency?

   A   Radio waves

   B   Ultraviolet

   C   Gamma rays

   D   Microwaves

(1 mark)

2   The speed of electromagnetic waves in a vacuum is approximately 300 000 000 m/s.
Use this figure to calculate the wavelength of a 100 MHz radio signal.

   A   3 m

   B   300 m

   C   30 000 m

   D   3 000 000 m

(1 mark)

3   Look at this displacement-time graph for a water wave.

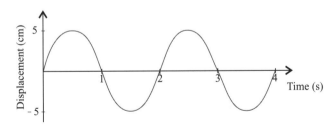

   (a)   What is the amplitude of this wave?

(1 mark)

   (b)   Calculate the frequency of the wave.

(1 mark)

   (c)   If the frequency of the wave doubles but its speed stays the same, what will happen to its wavelength?

(1 mark)

Exam Questions

4   The diagram shows a light ray entering and leaving a glass block.

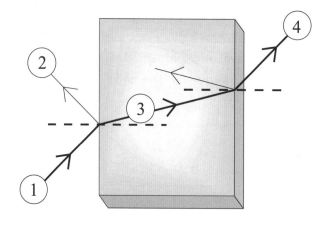

Match **A**, **B**, **C** and **D** to labels **1-4** on the diagram.

A   A reflected ray

B   A refracted ray

C   An emergent ray

D   An incident ray

(4 marks)

5   The diagram shows a light ray entering an optical fibre.

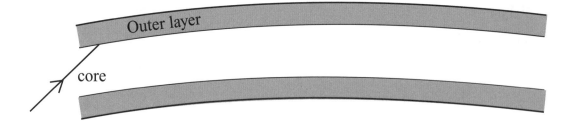

(a)   Sketch the path taken by the light ray.

(1 mark)

(b)   Explain why the light ray follows this path.

(1 mark)

Dangers of EM Radiation

Electromagnetic radiation can be dangerous. (So it'll probably be banned one of these days... sigh.)

Some radiations are **more harmful** than others

When EM radiation enters living tissue — like you — it's often harmless, but sometimes it creates havoc.

1) Some EM radiation mostly passes through soft tissue without being absorbed — e.g. radio waves.

2) Other types of radiation are absorbed and cause heating of the cells — e.g. microwaves.

3) Some radiations cause cancerous changes in living cells — e.g. UV can cause skin cancer.

4) Some types of EM radiation can actually destroy cells — as in 'radiation sickness' after a nuclear accident.

Higher frequency EM radiation is usually **more dangerous**

1) As far as we know, radio waves are pretty harmless.

2) This is because the energy of any electromagnetic wave is directly proportional to its frequency.

3) Visible light isn't harmful unless it's really bright. People who work with powerful lasers (very intense light beams) need to wear eye protection.

4) Infrared can cause burns or heatstroke (when the body overheats) — but they're easily avoidable risks.

In general, waves with lower frequencies (like radio) are less harmful than high frequency waves like X-rays and gamma rays.

Higher frequency waves have more energy. And it's the energy of a wave that does the damage.

Microwaves — may or may not be harmful

1) Some wavelengths of microwaves are absorbed by water molecules and heat them up. If the water in question happens to be in your cells, you might start to cook.

2) Mobile phone networks use microwaves, and some people think that using your mobile a lot, or living near a mast, could damage your health. There isn't any conclusive proof either way yet.

In general, the more energy a wave has, the more damage it can do

If you can remember the electromagnetic spectrum, this page should be a doddle. Just think about the two ends of the spectrum — would you rather have some tuneful radio waves or scary gamma rays? I'd go for the radio waves any day — the dangers of the other types of radiation fit nicely in between.

Dangers of EM Radiation

Information about the dangers of too much Sun is everywhere nowadays, but getting a tan can't be that bad for you... can it?

Ultraviolet radiation can cause skin cancer

1) If you spend a lot of time in the sun, you'll get a tan and maybe sunburn (with attractive peeling).

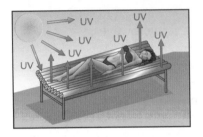

2) The more time you spend in the sun, the more chance you also have of getting skin cancer. This is because the Sun's rays include ultraviolet radiation (UV) which damages the DNA in your cells.

3) Dark skin gives some protection against UV rays — it absorbs more UV radiation, stopping it from reaching the more vulnerable tissues deeper in the body.

4) Everyone should protect themselves from overexposure to the Sun, but if you're pale skinned, you need to take extra care, and use a sunscreen with a higher Sun Protection Factor (SPF).

An SPF of 15 means you can spend 15 times as long in the sun as you otherwise could without burning (if you keep reapplying the sunscreen).

5) The gas inside fluorescent tubes (often used for kitchen and office lighting) emits UV radiation. Special coatings are used on lamps to absorb the UV and emit visible light instead.

The ozone layer protects us from UV radiation

The ozone layer absorbs some UV rays from the Sun — so it reduces the amount of UV radiation reaching the Earth's surface. Recently, the ozone layer has got thinner because of pollution from CFCs.

- CFCs are gases which react with ozone molecules and break them up. This depletion of the ozone layer allows more UV rays to reach us at the surface of the Earth.

- We used to use CFCs all the time — e.g. in hairsprays and in the coolant for fridges — they're now banned or restricted because of their environmental impact.

Staying in the sun for too long can cause cancer

There's no point being paranoid — a little bit of sunshine won't kill you (in fact it might do you good). But don't be daft... getting cancer from sunbathing for hours on end is just stupid. As for CFCs — that's chlorofluorocarbons by the way — they're mostly banned now but the ozone layer is still being damaged.

Warm-Up and Exam Questions

You must be getting used to the routine by now — the warm-up questions get you, well, warmed up, and the exam questions give you some idea of what you'll have to cope with on the day.

Warm-Up Questions

1) Explain how high frequency electromagnetic radiation can damage cells.
2) Why can it be dangerous for living cells to absorb infrared radiation?
3) Why is the use of CFCs now restricted?
4) Which type of radiation does the ozone layer absorb?

Exam Questions

1   The radiation used in mobile phone networks

   A   is definitely safe

   B   has a higher frequency than visible light

   C   may cause damage to cells

   D   is known to cause health problems

(1 mark)

2   Match the words **A**, **B**, **C** and **D** with the spaces **1-4** in the sentences.

   A   Visible light

   B   Radio

   C   Infrared

   D   Ultraviolet

Using sunscreen will give you some protection from ...**1**... radiation emitted by the Sun.

Grills use ...**2**... radiation to cook food.

...**3**... is not usually dangerous unless it's very bright e.g. lasers.

...**4**... waves are not known to have any ill effects on the body.

(4 marks)

3   (a) Give one hazard and one practical use of

     (i)   microwave radiation

(2 marks)

     (ii) gamma radiation

(2 marks)

   (b) Give one way in which you can reduce your exposure to UV radiation from the Sun.

(1 mark)

Uses of Waves

Radio waves have <u>long wavelengths</u> and <u>low frequencies</u> — making them great for communications.

Radio waves are used mainly for communications

1) <u>Radio waves</u> (EM radiation with wavelengths longer than about 10 cm) and some <u>microwaves</u> are good at transferring information <u>long distances</u>.

2) This is partly because they don't get <u>absorbed</u> by the Earth's atmosphere as much as waves in the <u>middle</u> of the EM spectrum (like heat, for example), or those at the high-frequency end of the spectrum (e.g. gamma rays or X-rays — though these would be too dangerous to use anyway).

3) But the different wavelengths have different properties, and are used in different ways...

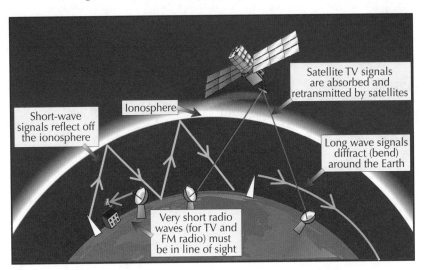

- <u>Long-wave radio</u> (wavelengths of <u>1 – 10 km</u>) can be transmitted from London, say, and received halfway round the world, because long wavelengths <u>diffract</u> around the curved surface of the Earth.

- The radio waves used for <u>TV and FM radio</u> transmissions, and the microwaves used for mobile phone communications, have very short wavelengths (10 cm – 10 m). These signals don't diffract much — to get reception, you must be in <u>direct sight of the transmitter</u>. This is why mobile phone transmitters are positioned on <u>hill tops</u> and fairly <u>close to one another</u>.

  Microwaves are used to carry satellite TV broadcasts or satellite phone calls...

  1) A <u>transmitter</u> on Earth sends the signal up into space...

  2) ...where it's picked up by the <u>satellite receiver dish</u> orbiting thousands of kilometres above the Earth. The satellite transmits the signal <u>back to Earth</u>...

  3) ...where it's picked up by a receiving <u>satellite dish</u>.

- <u>Short-wave radio</u> signals (wavelengths of about <u>10 m – 100 m</u>) can, like long wave, be received at <u>long distances</u> from the transmitter. That's because they're <u>reflected</u> from the <u>ionosphere</u> — an <u>electrically charged layer</u> in the Earth's upper atmosphere.

That's why hilly areas can get French radio better than Radio 4

Radio waves can be long, medium or short wave — but don't be fooled, short-wave radio waves still have a longer wavelength than other sorts of electromagnetic waves. There's short and there's short.

Uses of Waves

You might take one look at this page and think it all looks fairly obvious — microwaves, used in microwave ovens and X-rays in X-ray machines, surely not. Ok, you probably knew that bit already, but this page does have lots of stuff about how they work, which isn't all that obvious, so don't miss it out.

Microwaves in **ovens** are **absorbed** by **water molecules**

1) The microwaves used in <u>microwave ovens</u> have a <u>different wavelength</u> to those used in communication.

2) These microwaves are actually <u>absorbed</u> by the water molecules in the food. They penetrate a few centimetres into the food before being <u>absorbed</u> by <u>water molecules</u>.

3) The energy is then <u>conducted</u> or <u>convected</u> to other parts.

4) If microwaves are absorbed by molecules in living tissue, <u>cells</u> may be <u>burned</u> or killed.

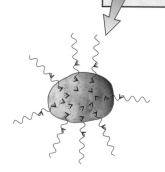

X-Rays are used to identify fractures

1) <u>Radiographers</u> in <u>hospitals</u> take <u>X-ray photographs</u> of people to see if they have any <u>broken bones</u>.

2) X-rays pass <u>easily through flesh</u> but not so easily through <u>denser material</u> like <u>bones</u> or <u>metal</u>. So it's the amount of radiation that's <u>absorbed</u> (or <u>not absorbed</u>) that gives you an X-ray image.

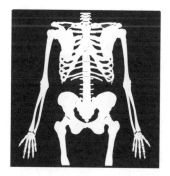

The <u>brighter bits</u> are where <u>fewer X-rays</u> get through. This is a <u>negative image</u>. The plate starts off <u>all white</u>.

3) X-rays can cause <u>cancer</u> (see page 57), so radiographers wear <u>lead aprons</u> and stand behind a <u>lead screen</u> or <u>leave the room</u> to keep their <u>exposure</u> to X-rays to a <u>minimum</u>.

Microwaves can be dangerous if they're absorbed by living tissue

X-rays are really useful for seeing inside people's bodies — but they can also cause cancer. So should we still use them? There's no right or wrong answer to questions like this — the best thing seems to be to only use x-rays when they're really needed and to try and protect the people who work with them.

Uses of Waves

You've seen how electromagnetic waves can be dangerous, but they can also be really useful...

Infrared radiation can be used to monitor temperature

1) Infrared radiation (or IR) is also known as heat radiation. It's given out by all hot objects — and the hotter the object, the more IR radiation it gives out.

2) This means infrared can be used to monitor temperatures (see page 17).

3) Infrared is also detected by night-vision equipment:

Heat radiation is given off by all objects, even in the dark of night. The equipment turns it into an electrical signal, which is displayed on a screen as a picture.

Prenatal scanning uses ultrasound, not EM Waves

Ultrasound is made up of high frequency sound waves. As these waves hit different media some of the waves are partially reflected. These reflected waves are processed by computer to produce a video image of the foetus. No one knows for sure whether ultrasound is safe in all cases, but X-rays would definitely be dangerous to the foetus.

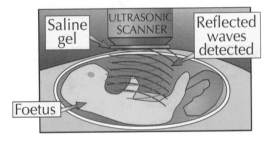

ADVANTAGES

1) As far as we know, it's safe.

2) It can show that the baby's alive and developing normally, determine its sex and show if it's likely to have Down's syndrome.

3) Ultrasound equipment is usually cheap and portable.

DISADVANTAGES

1) We can't be totally sure it's safe.

2) There are ethical issues — the parents may be more likely to want an abortion if the foetus is 'the wrong sex' or if it has Down's syndrome. Is it right for parents to have this choice?

It's a shame facts aren't as easily absorbed as infrared radiation

I know there's loads of uses of waves, but you really do need to know about them. Try writing a table with the type of wave, a use of it and a brief outline of how it works — it'll be a barrel of fun. Ok, maybe not, but it will be useful — you can even go back later and read it to get it in your head.

Uses of Waves

Yes, there's more — waves are just great aren't they.

Iris scanning uses light to make a picture

1) Each eye's iris has a unique pattern — not even your own eyes have the same pattern as each other. This means that iris patterns can be used in security checks to prove a person's identity.

2) This is done using an iris scanner — which is basically like a camera. Light is reflected off your iris to make a picture. A computer then analyses the patterns in your iris to check your identity.

ADVANTAGES

1) There's very little chance of mistaking one iris code for another.

2) Your iris won't (usually) change during your life.

3) Iris recognition systems have more reference points for comparison than fingerprints.

4) Iris scanning is quick and easy to carry out.

DISADVANTAGES

1) Eye injuries or surgery (e.g. cataract removal) can occasionally change your iris pattern.

2) Some people feel their personal freedom is threatened: who's holding our data, and why?

3) If iris data is assumed to be 'unfakeable', will genuine mistakes or stolen identity be believed?

CD players use lasers to read information

1) The surface of a CD has a pattern of shallow pits (and higher areas called 'lands') cut into it.

2) A laser shone onto the CD is reflected from the shiny surface as it spins around in the player.

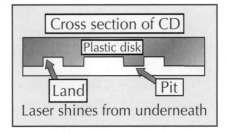

Cross section of CD

Plastic disk

Land   Pit

Laser shines from underneath

3) The beam is reflected from a land and a pit slightly differently — and this difference can be picked up by a light sensor. These differences in reflected signals can then be changed into an electrical signal.

4) An amplifier and a loudspeaker then convert the electrical signal into sound of the right pitch (frequency) and loudness.

Look into my eyes and tell me who I am

These are the last two uses of waves, I promise. Don't forget to add them to your table — you might want to add an advantages and disadvantages column too, so you can learn all about that.

Warm-Up and Exam Questions

There's a knack to passing exams — applying all the facts you've got stored in your brain to get as many marks as possible. These exam questions will help you to practise that.

Warm-Up Questions

1) Suggest two reasons why radio waves are suitable for using for communication.
2) Explain how microwaves can heat food.
3) What is the difference between the microwaves used for communications and those used to heat food?
4) Which type of wave is 'bounced back' by the ionosphere?
5) Describe how CD players use lasers to convert the data on the CD into electrical signals.

Exam Questions

1   Which type of electromagnetic radiation can be detected by thermal imaging cameras?

    A   Radio waves

    B   Microwaves

    C   Infrared

    D   X-rays

(1 mark)

2   The Government is planning to introduce identity cards containing biometric information such as iris patterns.

    (a) Describe briefly how an iris scanner creates an image of an iris pattern.

(1 mark)

    (b) Give one reason why checking iris patterns is a good way of identifying people.

(1 mark)

    (c) Give one disadvantage of using iris recognition as an identification method.

(1 mark)

3   Jamie has an X-ray taken to see if he has broken his arm.

    (a) Why are X-rays not very useful for looking at soft tissue injuries?

(1 mark)

    (b) Give two ways in which radiographers can minimise their exposure to X-rays.

(2 marks)

4   Ultrasound is used to produce an image of the foetus in prenatal scans.

    Give one advantage and one disadvantage of carrying out prenatal scans.

(2 marks)

Analogue and Digital Signals

Digital technology is gradually taking over. By 2012, you won't be able to watch TV unless you've got a digital version — that's when the Government's planning to switch off the last analogue signal.

Information *is* converted *into* signals

Information is being transmitted everywhere all the time.

1) Information, such as sounds and pictures, is converted into electrical signals before it's transmitted.

2) It's then sent long distances down telephone wires or carried on EM waves.

Analogue *signals* vary...

1) The amplitude and frequency of an analogue signal vary continuously. An analogue signal can take any value in a particular range.

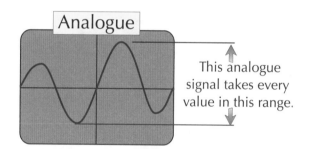

This analogue signal takes every value in this range.

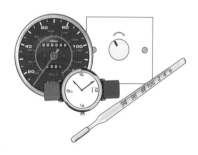

2) Dimmer switches, thermometers, speedometers and old-fashioned watches are all analogue devices.

...but digital's *either* on *or* off

1) Digital signals can only take two values (the two values sometimes get different names, but the key thing is that there are only two of them): on or off, true or false, 0 or 1...

Digital

This digital signal only takes these two values.

2) On/off switches and the displays on digital clocks and meters are all digital devices.

Analogue and Digital Signals

It's all very well being able to transmit signals, but you need to be able to makes sense of them when you receive them too, or it's all a bit of a waste of time.

Signals have to be amplified

Both digital and analogue signals <u>weaken</u> as they travel, so they may need to be <u>amplified</u> along their route.

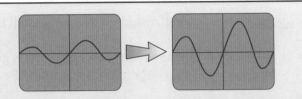

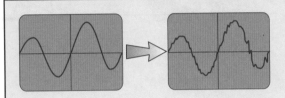

They also pick up <u>interference</u> or <u>noise</u> from <u>electrical disturbances</u> or <u>other signals</u>.

Digital signals are far better quality

1) <u>Noise</u> is less of a problem with <u>digital</u> signals.
   If you receive a noisy digital signal, it's pretty obvious what it's supposed to be.

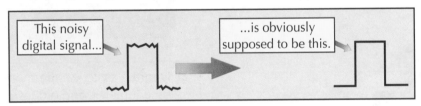

This noisy digital signal... ...is obviously supposed to be this.

2) But if you receive a noisy <u>analogue</u> signal, it's difficult to know what the <u>original</u> signal would have looked like. And if you amplify a noisy analogue signal, you amplify the noise as well.

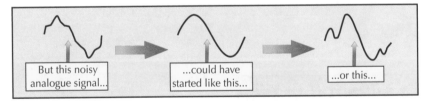

But this noisy analogue signal... ...could have started like this... ...or this...

3) This is why digital signals are much <u>higher quality</u> — the information received is the <u>same</u> as the original.

4) Digital signals are also easy to <u>process</u> using <u>computers</u>, since computers are digital devices too.

5) And another advantage of digital technology is that you can transmit <u>several signals at once</u> using just one cable or EM wave — so you can send <u>more information</u> (in a given time) than using analogue signals.

Remember: analogue varies but digital's either on or off

Digital signals are great — unless you live in a part of the country which currently has poor reception of digital broadcasts, in which case you get <u>no benefit at all</u>. This is because if you don't get spot-on reception of digital signals in your area, you won't get a grainy but watchable picture (like with analogue signals) — you'll get nothing at all. Except snow.

Seismic Waves

Seismic waves are completely different from EM waves (although they're still waves, so all the 'normal wave stuff' still applies). Seismic waves are produced by earthquakes.

Seismic waves are used to investigate Earth's structure

1) When there's an earthquake somewhere, it produces seismic waves which travel out through the Earth. We detect these waves all over the surface of the planet using seismographs.

2) There are two different types of seismic waves you need to learn — P-waves and S-waves.

3) Seismologists work out the time it takes for the shock waves to reach each seismograph. They also note which parts of the Earth don't receive the shock waves at all.

P-waves are longitudinal

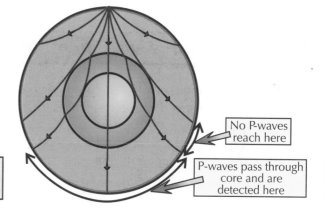

1) P-waves travel through solids and liquids.

No P-waves reach here

P-waves pass through core and are detected here

2) They travel faster than S-waves.

S-waves are transverse

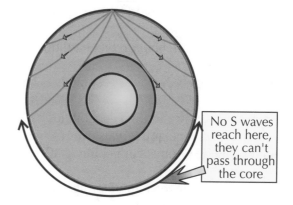

1) S-waves only travel through solids.

No S waves reach here, they can't pass through the core

2) They're slower than P-waves.

You need to know the differences between S-waves and P-waves

Seismic waves might not seem earth-shakingly exciting to you, but those examiners love to ask about the differences between the two sorts, so you should try and learn them. It's not as hard as you might think, just remember to keep all the s-words together — s-waves, tranSverSe, solids, slow. It doesn't work quite as well for p-waves — but if you know what s-waves are you should be able to work it out.

Seismic Waves

The **seismograph** results tell us **what's down there**

1) About <u>halfway through</u> the Earth, both types of wave <u>change direction</u> abruptly. This indicates that there's a <u>sudden change</u> in <u>properties</u> — as you go from the <u>mantle</u> to the <u>core</u>.

2) <u>S-waves</u> do travel through the <u>mantle</u>, which shows that it's <u>solid</u>. But the fact that <u>S-waves</u> are <u>not detected</u> in the core's <u>shadow</u> tells us that the <u>outer core</u> is <u>liquid</u> — <u>S</u> waves only pass through <u>S</u>olids.

3) It's also found that <u>P-waves</u> travel <u>slightly faster</u> through the <u>middle</u> of the core, which strongly suggests that there's a <u>solid inner core</u>.

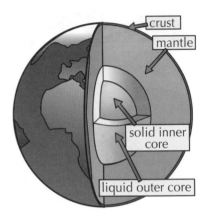

The waves **curve** with **increasing depth**

1) The <u>waves</u> change speed as the <u>properties</u> of the mantle and core change. This change in speed causes the waves to change direction — which is <u>refraction</u>, of course.

2) Most of the time the waves change speed <u>gradually</u>, resulting in a <u>curved path</u>. But when the properties change <u>suddenly</u>, the wave speed changes abruptly, and the path has a <u>kink</u>.

It's **difficult** to **predict earthquakes** and **tsunami waves**

<u>Some countries</u> are particularly <u>susceptible</u> to <u>earthquakes</u>, which can also sometimes <u>cause tsunami waves</u>. Both can be <u>extremely destructive</u>, especially in areas where the <u>housing</u> isn't built to <u>withstand</u> them. So it would be <u>very useful</u> to be able to <u>predict</u> when they're <u>likely to hit</u>.

1) Seismic waves can help predict earthquakes — a 'big one' usually happens after smaller 'foreshocks'.

2) But predicting earthquakes accurately is <u>hard</u>. Scientists don't agree on which method works best.

3) For example, you can use <u>probabilities</u> based on <u>previous occurrences</u> — e.g. if a city has had an earthquake at regular intervals over the last century, that <u>trend may continue</u>. This method isn't necessarily dead-on accurate, but it still gives the area <u>time to prepare</u>, just in case.

4) Other methods involve <u>monitoring</u> various factors such as the level of <u>groundwater</u>, <u>foreshocks</u>, the <u>emission of radon gas</u>, and even, in some countries, changes in <u>animal behaviour</u>. These methods have been <u>successful</u>, but have also <u>missed</u> some biggies.

Seismic waves are a good way of investigating the Earth's structure
Scientists are still doing lots of research into earthquake prediction — and disasters like the Indian Ocean tsunami in 2004 and the Kashmir earthquake in 2005 have really raised its profile of late.

Warm-Up and Exam Questions

Learning facts and practising exam questions is the only recipe for sure-fire success.
That's what the questions on this page are all about. All you have to do — is do them.

Warm-Up Questions

1) Describe the difference between analogue and digital signals.
2) Which type of seismic wave travels faster — P-waves or S-waves?
3) Which type of seismic wave is longitudinal?
4) Why do seismic waves curve as they travel through the Earth?
5) Give one way in which scientists try to predict earthquakes.

Exam Questions

1   Communication signals pick up unwanted additional signals as they travel.
     This unwanted part of the signal is called

   A   analogue

   B   amplitude

   C   digital

   D   noise

   (1 mark)

2   When comparing P-waves and S-waves travelling through the Earth, which of the
     following statements is **not** true?

   A   They both travel through liquids.

   B   They both travel through solids.

   C   P-waves are faster than S-waves.

   D   They are both caused by earthquakes.

   (1 mark)

3   Rick is sick of not being able to get a clear signal on his analogue radio.
     He decides to replace it with a digital radio.

   (a)   What values can digital signals take?

   (1 mark)

   (b)   Explain why both digital and analogue radio signals need to be amplified when they
         reach his radio.

   (1 mark)

   (c)   Give one reason why Rick will be able to hear the songs on the radio more clearly
         with his digital radio than with his analogue radio.

   (1 mark)

Revision Summary for Section Two

Try these lovely questions. Go on — you know you want to. It'll be nice.

1) Explain what current, voltage and resistance are in an electric circuit.

2) Do batteries produce AC or DC current?

3) Sketch typical voltage-current graphs for: a) a fixed value resistor, and b) a filament lamp.
Explain the shape of each graph.

4)* Calculate the resistance of a wire if the voltage across it is 12 V and the current through it is 2.5 A.

5) Describe how the resistance of an LDR varies with light intensity. Give an application of an LDR.

6) a) Describe how you can create a current in a coil of wire using a magnet.
b) What are three factors that affect the size of the voltage that you get this way?

7) Explain how a generator works.

8) Draw a diagram showing how a typical power station works, and explain what happens at each stage.

9) Explain why a very high electrical voltage is used to transmit electricity in the National Grid.

10) What's the name of the type of transformer that increases voltage? Where in the National Grid
are these used?

11)*a) How many units of electricity (in kWh) would a kettle of power 2500 W use in 2 minutes?
b) How much would that cost, if one unit of electricity costs 12p?

12) Draw a diagram of a wave and label a crest and a trough, and the wavelength and amplitude.

13) Sketch the EM spectrum with all the details you've learned. Put the lowest frequency waves first.

14) Electromagnetic waves don't carry any matter. What <u>do</u> they carry?

15) What aspect of EM waves determines their differing properties?

16)*Find the speed of a wave with frequency 50 kHz and wavelength 0.3 cm.

17) Describe what can happen to a wave's direction when it meets an obstacle.

18) Draw a diagram showing the diffraction of a wave as it passes through: a) a small gap, b) a big gap.

19) Explain what is meant by: a) refraction b) total internal reflection.

20)*In which of the cases A to D below would the ray of light be totally internally reflected?
(The critical angle for glass is approximately 42°.)

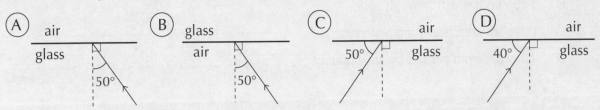

21) What has led to a thinning of the ozone layer? Why is this a problem for humans?

22) Explain why sending data by optical fibre might be better than broadcasting it as a radio signal.

23) Describe the main <u>known</u> dangers of microwaves, infrared, visible light, UV and X-rays.

24) Describe the different ways that short, medium and long-wave radio signals can travel long distances.

25) Describe two uses of microwaves.

26) Explain how EM waves are used in: a) prenatal scanning, b) iris scanning, c) CD players.

27) Draw diagrams illustrating analogue and digital signals. What advantages do digital signals have?

28) How do P-waves and S-waves differ regarding: a) what type of wave they are, b) their speed, c) what
they can travel through?

* Answers on page 272.

Radioactivity

Nuclear radiation is different from EM radiation. So you do need to read these next few pages. Sorry.

Nuclei contain protons and neutrons

The nucleus contains protons and neutrons. It makes up most of the mass of the atom, but takes up virtually no space — it's tiny.

The electrons are negatively charged and really really small.

They whizz around the outside of the atom. Their paths take up a lot of space, giving the atom its overall size (though it's mostly empty space).

Isotopes are atoms with different numbers of neutrons

1) Many elements have a few different isotopes. Isotopes are atoms with the same number of protons but a different number of neutrons.

2) E.g. there are two common isotopes of carbon. The carbon-14 isotope has two more neutrons than 'normal' carbon (carbon-12).

3) Usually each element only has one or two stable isotopes — like carbon-12. The other isotopes tend to be radioactive — the nucleus is unstable, so it decays (breaks down) and emits radiation. Carbon-14 is an unstable isotope of carbon.

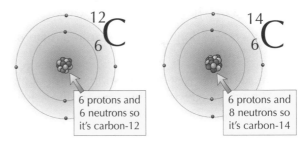

$^{12}_{6}C$ $^{14}_{6}C$

6 protons and 6 neutrons so it's carbon-12

6 protons and 8 neutrons so it's carbon-14

Radioactive decay is a random process

1) The nuclei of unstable isotopes break down at random. If you have 100 unstable nuclei, you can't say when any one of them is going to decay, and you can't do anything to make a decay happen.

2) Each nucleus just decays quite spontaneously in its own good time. It's completely unaffected by physical conditions like temperature or any sort of chemical bonding etc.

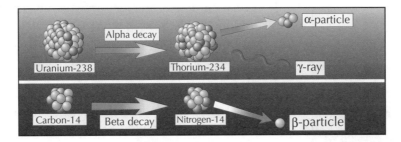

Uranium-238   Alpha decay   Thorium-234   α-particle   γ-ray

Carbon-14   Beta decay   Nitrogen-14   β-particle

3) When the nucleus does decay it spits out one or more of the three types of radiation — alpha, beta and gamma (see pages 72-73).

4) In the process, the nucleus often changes into a new element.

Protons determine the element, neutrons determine the isotope...

This isotope business can be a bit confusing at first, as you can have different isotopes which are all the same element. Remember, it's the number of protons which decides what element it is, then the number of neutrons decides what isotope of that element it is.

Radioactivity

There are three types of radiation — alpha (α), beta (β) (on this page) and gamma (γ) (on the next page). You need to remember <u>what</u> they are, how well they <u>penetrate</u> materials (including air), and their <u>ionising</u> power.

Nuclear radiation causes ionisation

1) Nuclear radiation causes <u>ionisation</u> by <u>bashing into atoms</u> and <u>knocking electrons off</u> them. Atoms (with <u>no overall charge</u>) are turned into <u>ions</u> (which are <u>charged</u>) — hence the term "ionisation".

2) There's a pattern: the <u>further</u> the radiation can <u>penetrate</u> before hitting an atom and getting stopped, the <u>less damage</u> it will do along the way and so the <u>less ionising</u> it is.

Alpha particles are helium nuclei $\Rightarrow$

1) Alpha particles are made up of <u>2 protons and 2 neutrons</u> — they're <u>big</u>, <u>heavy</u> and <u>slow-moving</u>.

2) They therefore <u>don't penetrate</u> far into materials but are <u>stopped quickly</u>.

3) Because of their size they <u>bash into a lot of atoms</u> and <u>knock electrons off</u> them before they slow down, which creates lots of ions.

4) Because they're electrically <u>charged</u> (with a positive charge), alpha particles are <u>deflected</u> (their <u>direction changes</u>) by <u>electric</u> and <u>magnetic fields</u>.

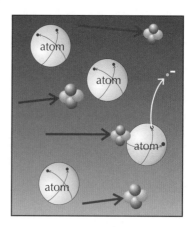

Beta particles are electrons $\Rightarrow$

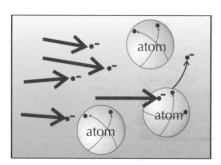

1) A beta particle is an <u>electron</u> which has been emitted from the <u>nucleus</u> of an atom when a <u>neutron</u> turns into a <u>proton</u> and an <u>electron</u>. So for every β-particle emitted, the number of <u>protons</u> in the nucleus increases by 1.

2) They move <u>quite fast</u> and they are <u>quite small</u>.

3) They <u>penetrate moderately</u> before colliding and are <u>moderately ionising</u> too.

4) Because they're <u>charged</u> (negatively), beta particles are <u>deflected</u> by electric and magnetic fields.

Learning the types of radiation is as easy as α, β, γ

The symbols for alpha, beta and gamma radiation may look a little strange — but really they're just a, b and c written using the Greek alphabet. True it might be easier to use a, b and c, but the Greek letters have been used for so long now that it'd confuse more people than it would help, sorry.

Radioactivity

Gamma radiation is quite different from alpha and beta radiation. Gamma rays are part of the electromagnetic spectrum — just like light and radio waves.

Gamma rays are very short wavelength EM waves

1) In a way, gamma rays are the opposite of alpha particles. They have no mass — they're just energy (in the form of an EM wave — see page 50).

2) They penetrate a long way into materials without being stopped.

3) This means they are weakly ionising because they tend to pass through rather than collide with atoms. But eventually they hit something and do damage.

4) Gamma rays have no charge, so they're not deflected by electric or magnetic fields.

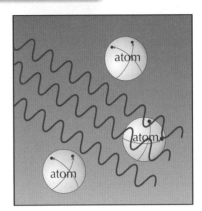

You can identify the type by what blocks it

Make sure you know what it takes to block each of the three types of radiation:

Alpha particles are blocked by paper, skin or a few centimetres of air.

Beta particles are stopped by thin metal.

Gamma rays are blocked by thick lead or very thick concrete.

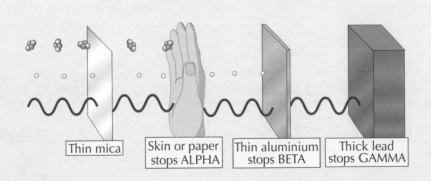

Thin mica

Skin or paper stops ALPHA

Thin aluminium stops BETA

Thick lead stops GAMMA

So if radiation can penetrate paper it could be beta or gamma — you'd have to test it with a metal, say, to find out which.

Remember — alpha penetrates least, gamma penetrates most

Remember: alpha's big, slow and clumsy — always knocking into things. Beta's lightweight and fast, and gamma weighs nothing and moves super-fast. Practise with this: if it gets through paper and is deflected by a magnetic field, it must be _____ radiation. (Answer on page 272.)

Background Radiation

We're constantly exposed to <u>very low levels</u> of radiation — and all without us noticing.

Background radiation comes from *many sources*

<u>Background radiation</u> comes from:

1) Radioactivity of naturally occurring <u>unstable isotopes</u> which are <u>all around us</u> — in the <u>air</u>, in <u>food</u>, in <u>building materials</u> and in the <u>rocks</u> under our feet.

2) Radiation from <u>space</u>, which is known as <u>cosmic rays</u>. These come mostly from the <u>Sun</u>. Luckily, the Earth's <u>atmosphere protects</u> us from much of this radiation. The Earth's <u>magnetic field</u> also deflects cosmic rays away from Earth.

3) Radiation due to <u>human activity</u>, e.g. <u>fallout</u> from <u>nuclear explosions</u> or <u>dumped nuclear waste</u>. But this represents a <u>tiny</u> proportion of the total background radiation.

The <u>relative proportions</u> of background radiation are:

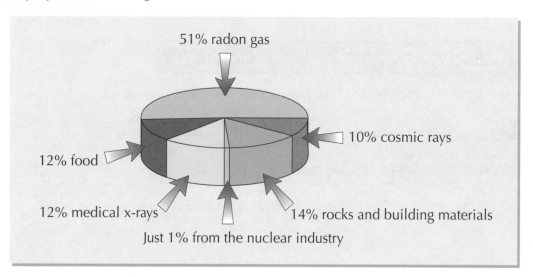

51% radon gas

10% cosmic rays

12% food

12% medical x-rays

14% rocks and building materials

Just 1% from the nuclear industry

The level of **background radiation** *depends on* **where you are**

1) At <u>high altitudes</u> (e.g. in <u>jet planes</u>) it <u>increases</u> because of more exposure to <u>cosmic rays</u>. That means commercial pilots have an increased risk of getting some types of cancer.

2) <u>Underground in mines</u>, etc. it increases because of the <u>rocks</u> all around.

3) Certain <u>underground rocks</u> (e.g. granite) can cause higher levels at the <u>surface</u>, especially if they release <u>radioactive radon gas</u>, which tends to get <u>trapped inside people's houses</u>.

Background radiation is everywhere

Background radiation was first discovered <u>accidentally</u>. Scientists were trying to work out which materials were radioactive, and couldn't understand why their reader still showed radioactivity, when there was <u>no material</u> being tested. They realised it must be natural background radiation.

Background Radiation

Radon gas accounts for over half of the <u>background radiation</u> on Earth.

*Exposure to **radon gas** causes **lung cancer***

1) Studies have shown that exposure to <u>high doses</u> of radon gas can cause <u>lung cancer</u> — and the <u>greater</u> the radon concentration, the <u>higher the risk</u>.

2) Some medical professionals reckon that about <u>1 in 20</u> deaths from <u>lung cancer</u> (about 2000 per year) are caused by radon exposure.

3) Evidence suggests that the risk of developing lung cancer from radon is <u>much greater</u> for <u>smokers</u> compared to nonsmokers.

4) The scientific community is a bit divided on the effects of <u>lower doses</u>, and there's still a lot of debate over what the highest safe(ish) concentration is.

*Radon gas concentration varies depends on **rock type***

The <u>radon concentration</u> in people's houses <u>varies widely</u> across the UK, depending on what type of <u>rock</u> the house is built on.

Level of radiation from rocks

Ventilation systems reduce radon concentration

1) <u>New houses</u> in areas where high levels of radon gas might occur must be designed with good <u>ventilation systems</u>. These reduce the concentration of radon in the living space.

2) In <u>existing houses</u>, the Government recommends that ventilation systems are put in wherever the radon concentration is higher than a certain level.

Radon gas is the largest source of background radiation

Radon gas is the subject of <u>scientific debate</u> — no one knows quite how much is too much. It's tricky to study because you can't accurately measure how much a person has been exposed to over their life. The best thing you can do is <u>not smoke</u>, as smoking increases your chance of developing lung cancer.

Half-Life

Half-life is a measure of the time it takes for a radioactive substance to decay.

The radioactivity of a sample always decreases over time

1) This is pretty obvious when you think about it. Each time a decay happens and an alpha, beta or gamma is given out, it means one more radioactive nucleus has disappeared.

2) Obviously, as the unstable nuclei all steadily disappear, the activity as a whole will decrease. So the older a sample becomes, the less radiation it will emit.

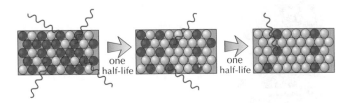

3) How quickly the activity decreases varies a lot. For some isotopes it takes just a few hours before nearly all the unstable nuclei have decayed. Others last for millions of years.

4) The problem with trying to measure this is that the activity never reaches zero, which is why we have to use the idea of half-life to measure how quickly the activity drops off.

5) Learn this important definition of half-life:

> ## HALF-LIFE is the TIME TAKEN for HALF of the NUCLEI now present to DECAY

Another useful definition of half-life is: "The time taken for the activity (or count rate) to fall by half".

6) A short half-life means the activity falls quickly, because lots of the nuclei decay quickly.

7) A long half-life means the activity falls more slowly because most of the nuclei don't decay for a long time — they just sit there, basically unstable, but kind of biding their time.

Do half-life questions step by step

A very simple example: The activity of a radioisotope is 640 cpm (counts per minute). Two hours later it has fallen to 40 cpm. Find the half-life of the sample.

You might also see radioactivity measured in becquerels (Bq). 1 Bq is 1 decay per second.

ANSWER: You must go through it in short simple steps like this:

Initial count:		after one half-life:		after two half-lives:		after three half-lives:		after four half-lives:
640	($\div 2$)→	320	($\div 2$)→	160	($\div 2$)→	80	($\div 2$)→	40

Notice the careful step-by-step method, which tells us it takes four half-lives for the activity to fall from 640 to 40. Hence two hours represents four half-lives, so the half-life is 30 minutes.

Half-Life

You can measure the <u>half-life</u> of a radioactive substance using a <u>G-M tube and counter</u>.

Measuring the half-life of a sample using a graph

1) This can <u>only be done</u> by taking <u>several readings</u> of <u>count rate</u>, usually using a <u>G-M tube and counter</u>.

2) The results can then be <u>plotted</u> as a <u>graph</u>, which will <u>always</u> be shaped like the one below.

3) The <u>half-life</u> is found from the graph, by finding the <u>time interval</u> on the <u>bottom axis</u> corresponding to a <u>halving</u> of the <u>activity</u> on the <u>vertical axis</u>.

Background radiation should be subtracted

<u>One trick</u> you need to know is about <u>background radiation</u>. This enters the G-M tube along with radiation from the sample giving a reading <u>higher</u> than the activity of the sample.

The solution is to measure the background count <u>first</u> and then <u>subtract it</u> from <u>every reading</u> you get, before plotting the results on the <u>graph</u>.

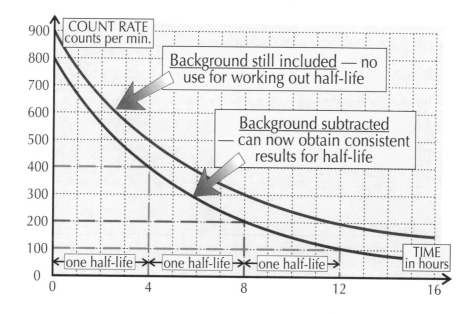

Realistically, the <u>only tricky bit</u> is actually <u>remembering</u> about background radiation in your <u>exam</u>, should they ask about it. They could also test the idea in a <u>calculation question</u>.

Half-lives always make curved graphs

The radioactivity of a sample always goes down over time, because every time a nucleus decays the sample gets smaller, so there are fewer nuclei left to decay. The result is that decay graphs always form nice curves like the one above. Plotting a decay graph is useful because it lets you calculate half-life.

Warm-Up and Exam Questions

There's no point in skimming through the section and glancing over the questions. Do the warm-up questions and go back over any bits you don't know. Then try the exam questions — without cheating.

Warm-Up Questions

1) What is meant by 'radioactive decay'?
2) Explain what isotopes are.
3) Name the three types of nuclear radiation.
4) Which type of nuclear radiation is also a type of electromagnetic radiation?
5) Give one source of background radiation.

Exam Questions

1   The diagram shows four different materials.

| thin mica | skin | aluminium | thick lead |
| 1 | 2 | 3 | 4 |

Match up the materials **1-4** with these descriptions.

A   Stops all types of nuclear radiation.

B   Doesn't stop any types of nuclear radiation.

C   Stops alpha and beta radiation.

D   Stops only alpha radiation.

(4 marks)

2   An alpha particle is

A   a proton

B   a neutron

C   a helium nucleus

D   an electromagnetic wave

(1 mark)

3   A sample of a highly ionising radioactive gas has a half-life of two minutes.

(a) What does 'half-life' mean?

(1 mark)

(b) What fraction of the radioactive atoms currently present will be left after four minutes?

(1 mark)

(c) When an atom of the gas decays, it releases an electron.
     What type of nuclear radiation does this gas emit?

(1 mark)

Dangers from Nuclear Radiation

Nuclear radiation can do nasty stuff to living cells so you have to be careful how you handle it. The effect of radiation on cells depends on the type of radiation and the size of the dose.

Radiation harms living cells...

1) Beta and gamma radiation can penetrate the skin and soft tissues to reach the delicate organs inside the body. This makes beta and gamma sources more hazardous than alpha when outside the body.

2) Alpha radiation can't penetrate the skin, but it's a different story when it gets inside your body (by swallowing or breathing it in, say) — alpha sources do all their damage in a very localised area.

3) Beta and gamma sources, however, are less dangerous inside the body — their radiation mostly passes straight out without doing much

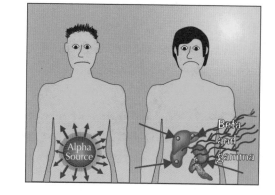

4) If radiation enters your body, it will collide with molecules in your cells.

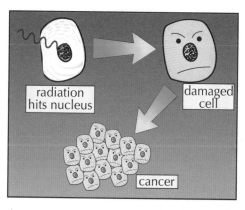

5) These collisions cause ionisation, which damages or destroys the molecules.

6) Lower doses tend to cause minor damage without killing the cell. This can give rise to mutant cells which divide uncontrollably — this is cancer.

7) Higher doses tend to kill cells completely, causing radiation sickness if a large part of your body is affected at the same time.

8) The extent of the harmful effects depends on how much exposure you have to the radiation, and its energy and penetration.

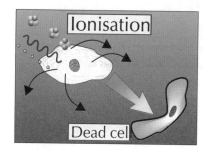

Nuclear radiation + living cells = cell damage, cell death or cancer

Most people could probably tell you that nuclear radiation is dangerous — what you need to know is what radiation can do to living cells and why the three types of radiation have different effects. Check out pages 72-73 if you're having trouble remembering the properties of the different radiation types.

Dangers from Nuclear Radiation

You've got to be really careful with anything radioactive — no mucking about.

You should **protect yourself** in the **laboratory**...

You should always act to <u>minimise</u> your exposure to radioactive sources.

1) <u>Never</u> allow <u>skin contact</u> with a source. Always handle sources with <u>tongs</u>.

2) Keep the source at <u>arm's length</u> to keep it <u>as far</u> from the body <u>as possible</u>.

3) Keep the source <u>pointing away</u> from the body and <u>avoid looking directly at it</u>.

4) <u>Always</u> keep the source in a <u>labelled lead box</u> and put it back in <u>as soon</u> as the experiment is <u>over</u>, to keep your <u>exposure time</u> short.

...and if you **work** with **nuclear radiation**

1) Industrial nuclear workers wear <u>full protective suits</u> to prevent <u>tiny radioactive particles</u> being <u>inhaled</u> or lodging <u>on the skin</u> or <u>under fingernails</u> etc.

2) <u>Lead-lined suits</u> and <u>lead/concrete barriers</u> and <u>thick lead screens</u> are used to prevent exposure to <u>gamma rays</u> from highly contaminated areas. (α and β radiation are stopped <u>much more easily</u>.)

3) Workers use <u>remote-controlled robot arms</u> to carry out tasks in highly radioactive areas.

Radiation's dangerous stuff — safety precautions are crucial

It's quite difficult to do research on how radiation affects humans. This is partly because it would be <u>unethical</u> to do <u>controlled experiments</u>, exposing people to huge doses of radiation just to see what happens. We rely mostly on studies of populations affected by <u>nuclear accidents</u> or nuclear <u>bombs</u>.

Uses of Nuclear Radiation

Nuclear radiation can be very <u>dangerous</u>. But it can be very <u>useful</u> too. Read on...

Alpha radiation is used in smoke detectors

1) Smoke detectors have a <u>weak</u> source of <u>alpha-radiation</u> close to <u>two electrodes</u>.

2) The radiation <u>ionises</u> the air, and a <u>current</u> flows between the electrodes.

3) But if there's a fire, the <u>smoke</u> <u>absorbs</u> the <u>radiation</u> — the <u>current stops</u> and the <u>alarm sounds</u>.

Beta radiation is used in tracers and thickness gauges

Medical tracers

1) Certain radioactive isotopes can be used as <u>tracers</u>. A tracer is <u>injected</u> into a patient (or <u>swallowed</u>) and its progress around the body is followed using an <u>external detector</u>. A computer converts the reading to a display showing where the <u>strongest reading</u> is coming from.

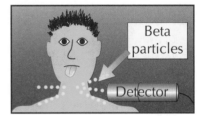

Beta particles

Detector

2) A well-known example is the use of <u>iodine-131</u>, which is absorbed by the <u>thyroid gland</u> just like normal iodine-127, but gives out <u>radiation</u> which can be <u>detected</u> to indicate whether the thyroid gland is <u>taking in iodine</u> as it should.

3) <u>All isotopes</u> which are taken <u>into the body</u> must be <u>BETA or GAMMA</u> emitters (never alpha), so that the radiation <u>passes out of the body</u> — and they should only last <u>a few hours</u>, so that the radioactivity inside the patient <u>quickly disappears</u> (i.e. they should have a <u>short half-life</u>).

Industry and thickness control

1) Tracers can be used in <u>industry</u> to find <u>leaks</u> and <u>blockages</u> in pipes.

2) <u>Beta radiation</u> is also used in <u>thickness control</u>.

radiation source

PAPER

β

hydraulic control

processor unit

detector

You direct radiation through the stuff being made (e.g. paper or cardboard), and put a detector on the other side, connected to a control unit. When the amount of <u>detected</u> radiation changes, it means the paper is coming out too thick or too thin, so the control unit adjusts the rollers to give the correct thickness.

The radioactive source used needs to have a fairly long half-life so it doesn't decay away <u>too quickly</u>.

It also needs to be a <u>beta</u> source, because then the paper will <u>partly block</u> the radiation. If it <u>all</u> goes through (or <u>none</u> of it does), then the reading <u>won't change</u> at all as the thickness changes.

Uses of Nuclear Radiation

Gamma radiation has medical and industrial uses

Treating cancer

1) High doses of <u>gamma rays</u> will kill living cells, so they can be used to treat <u>cancers</u>.

2) The gamma rays have to be <u>directed carefully</u> at the cancer, and at just the right <u>dosage</u> so as to kill the <u>cancer</u> cells <u>without</u> damaging too many <u>normal</u> cells.

Sterilising medical instruments

1) Gamma rays are also used to <u>sterilise</u> medical instruments — by <u>killing</u> all the microbes.

2) This is better than trying to <u>boil</u> plastic instruments, which might be damaged by high temperatures.

3) You need to use a <u>very strong</u> emitter of <u>gamma rays</u> with a <u>reasonably long</u> <u>half-life</u> so it doesn't need <u>replacing</u> too often.

Non-destructive testing

Several industries also use gamma radiation to do <u>non-destructive testing</u>.

For example, <u>airlines</u> can check the turbine blades of their jet engines by directing gamma rays at them — if too much radiation <u>gets through</u> the blade to the <u>detector</u> on the other side, they know the blade's <u>cracked</u> or there's a fault in the welding.

It's so much better to find this out before you take off than in midair.

The isotope you use depends on half-life and whether it's α, β or γ

Knowing the detail is important here. For instance, swallowing an alpha source as a medical tracer would be very foolish — alpha radiation would cause all sorts of chaos inside your body but couldn't be detected outside, making the whole thing pointless. So learn <u>what</u> each type's used for <u>and why</u>.

Radioactive Dating

Radioactive dating is used to work out the age of materials.

Radioactive dating of *rocks* and archaeological *specimens*

1) The discovery of radioactivity and the idea of <u>half-life</u> (see p76) gave scientists their <u>first opportunity</u> to <u>accurately</u> work out the <u>age</u> of <u>rocks</u>, <u>fossils</u> and <u>archaeological specimens</u>.

2) By measuring the <u>amount</u> of a <u>radioactive isotope</u> left in a sample, and knowing its <u>half-life</u>, you can work out <u>how long</u> the thing has been around.

3) <u>Igneous</u> rocks contain radioactive uranium which has a ridiculously <u>long half-life</u>. It eventually decays to become <u>stable</u> isotopes of <u>lead</u>, so the big clue to a rock sample's age is the <u>relative proportions</u> of uranium and lead isotopes.

4) Igneous rock also contains the radioisotope <u>potassium-40</u>. Its decay produces stable <u>argon gas</u> and sometimes this gets trapped in the rock. Then it's the same process again — finding the <u>relative proportions</u> of potassium-40 and argon to work out the age.

Carbon-14 calculations — *carbon dating*

1) <u>Carbon-14</u> makes up about 1/10 000 000 (one <u>ten-millionth</u>) of the carbon in the <u>air</u>. This level stays <u>fairly constant</u> in the <u>atmosphere</u>.

2) The same proportion of C-14 is also found in <u>living things</u>.

3) However, when they <u>die</u>, the C-14 is <u>trapped inside</u> the wood or wool or whatever, and it <u>gradually decays</u> with a <u>half-life</u> of <u>5730 years</u>.

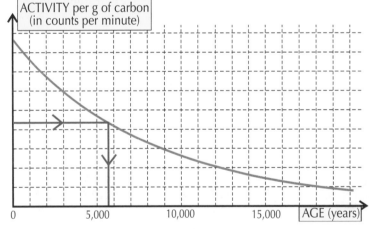

4) So, by measuring the <u>proportion</u> of C-14 found in some old <u>axe handle</u>, <u>burial shroud</u>, etc. you can calculate <u>how long ago</u> the item was <u>living material</u> using the known <u>half-life</u>.

<u>EXAMPLE:</u> An axe handle was found to contain 1 part in 40 000 000 C-14. How old is the axe?

<u>ANSWER:</u> The C-14 was originally <u>1 part in 10 000 000</u>. After <u>one half-life</u> it would be down to <u>1 part in 20 000 000</u>. After <u>two half-lives</u> it would be down to <u>1 part in 40 000 000</u>. Hence the axe handle is <u>two C-14 half-lives</u> old, i.e. 2 × 5730 = <u>11 460 years old</u>.

Radioactive Dating

Radioactive dating can only give you an <u>estimate</u> of age...

*The **results** from radioactive dating **aren't perfect***

<u>Scientific conclusions</u> are often <u>full of uncertainties</u>. This is because they're based on certain <u>assumptions</u> which may not always be true.

Carbon dating is based on the assumption that...

1) The <u>level of C-14</u> in the atmosphere has always been constant.

2) All living things take in the <u>same proportion</u> of their carbon as C-14.

3) Substances haven't been <u>contaminated</u> by a more recent source of carbon (i.e. after they've died).

...but in reality...

1) The level of C-14 <u>hasn't always been constant</u> — cosmic radiation, climate change and human activity have all had an effect. To account for this, scientists adjust their results using <u>calibration tables</u>.

2) Not all living things act as we <u>expect</u>. For example, some plants take up <u>less C-14</u> than expected — making them seem older than they really are.

3) Scientists can't be <u>100% sure</u> that a sample hasn't been contaminated.

*Measuring error also affects the **accuracy** of results*

The <u>proportion of C-14</u> measured in a sample is unlikely to be exact — either because of the <u>equipment</u> used or <u>human error</u>.

When conducting scientific research, scientists can use <u>technology</u> to increase the accuracy of results. Measurements of radioactivity can be made more accurately by attaching the counter to a <u>computer</u>. Instead of taking readings and plotting a graph by hand, the computer's software plots the graph for you.

To do carbon dating you need to know the half-life of carbon

Carbon dating was originally developed by a team of scientists lead by <u>Willard Libby</u> in the late 1940s. Libby's team was the first to measure the half-life of C-14. The value they came up with (<u>5568 ± 30 years</u>) is very close to the value accepted today — <u>5730 ± 40 years,</u> known as the <u>Cambridge half-life</u>.

Warm-Up and Exam Questions

Imagine if you opened up your exam paper and all the answers were already written in for you.
Hmm, well I'm afraid that won't happen, so the only way you'll do well is through some hard work now.

Warm-Up Questions

1) How do smoke detectors work?
2) Why is nuclear radiation dangerous?
3) Describe two precautions that should be taken when handling radioactive sources in the lab.
4) Give one way in which workers in nuclear power plants can be protected from radiation.
5) Give one example of a medical use of nuclear radiation.
6) Give three assumptions made when carrying out radiocarbon dating.

Exam Questions

1   Which type of nuclear radiation is the most dangerous inside the body?

   A   alpha

   B   beta

   C   gamma

   D   neutron

(1 mark)

2   Temperature-sensitive surgical instruments are sterilised by

   A   alpha radiation

   B   boiling

   C   beta radiation

   D   gamma radiation

(1 mark)

3   The diagram shows paper being made in a mill.

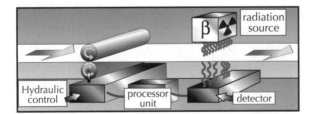

(a)   Describe how the thickness of the paper is controlled using beta radiation.

(2 marks)

(b)   Why isn't an alpha source used with this machinery?

(1 mark)

(c)   Why isn't gamma radiation used for this purpose?

(1 mark)

The Solar System

Our Solar System consists of a star (the Sun) and lots of stuff orbiting it in elongated circles.

There are **eight planets** in our Solar System

- Closest to the Sun are the <u>inner planets</u> — Mercury, Venus, Earth and Mars.
- Then the <u>asteroid belt</u> — see below.
- Then the <u>outer planets</u>, much further away — Jupiter, Saturn, Uranus, Neptune.

We used to say there was a ninth planet called Pluto, but scientists have decided it's too small to be counted as a planet. It's now called 'minor planet 134340 Pluto'— catchy.

You need to learn the <u>order</u> of the planets, which is made easier by using the little jollyism below:

Mercury,	Venus,	Earth,	Mars,	(Asteroids),	Jupiter,	Saturn,	Uranus,	Neptune
(Mad	Vampires	Eat	Mangos	And	Jump	Straight	Up	Noses)

Planets **reflect sunlight** and orbit the **Sun** in **ellipses**

1) You can <u>see</u> some planets with the <u>naked eye</u> — they look like <u>stars</u>, but they're <u>totally different</u>.

2) Stars are <u>huge</u>, very <u>hot</u> and very <u>far away</u>. They <u>give out</u> lots of <u>light</u> — which is why you can see them, even though they're very far away.

3) The planets are <u>smaller</u> and <u>nearer</u> and they just <u>reflect sunlight</u> falling on them.

4) Planets often have <u>moons</u> orbiting around them. Jupiter has at least 63 of 'em. We've just got one.

There's a **belt** of **asteroids** orbiting between **Mars** and **Jupiter**

1) When the Solar System formed, there were lots of leftovers — <u>bits of rock and rubble</u>. These included <u>asteroids</u>, which orbit the Sun in the <u>asteroid belt</u>, between the orbits of <u>Jupiter</u> and <u>Mars</u>. (They didn't get to form a planet because the large gravitational force of Jupiter kept interfering.)

2) Occasionally lumps of rock get <u>pushed out</u> of their <u>nice safe orbits</u> (e.g. due to a collision) — they might then enter the Earth's atmosphere...

3) ...where they're called <u>meteors</u>. As they pass through our <u>atmosphere</u> they usually <u>burn up</u> and we see them as <u>shooting stars</u>.

4) Sometimes, not all of the meteor burns up and part crashes into the <u>Earth's surface</u> as a <u>meteorite</u>.

The Solar System

Every now and then an asteroid or comet comes near the Earth — scary stuff. Luckily, the chances of an object hitting the Earth and being big enough to do serious damage is really low, phew.

Comets *orbit the Sun in very* **elliptical orbits**

1) <u>Comets</u> are balls of <u>ice</u> and <u>dust</u> which orbit the Sun in very <u>elongated ellipses</u> — taking them <u>very far away</u> and then back in <u>close</u> — which is when <u>we</u> see them.

2) The Sun is <u>near one end</u> of the orbit, not at its centre.

3) As a comet approaches the Sun, it speeds up. Also, its <u>ice melts</u>, leaving a <u>bright tail</u> of gas and debris which can be <u>millions of kilometres</u> long.

Near-Earth Objects *(NEOs) could* **collide** *with* **Earth**

1) Some asteroids and comets have orbits that pass <u>very close</u> to the orbit of the Earth. [They're called <u>Near-Earth Objects</u> (NEOs).]

2) Astronomers use <u>powerful telescopes</u> and <u>satellites</u> to search for and monitor NEOs. Then they can calculate an object's <u>trajectory</u> (the path it's going to take) and find out if it's heading for <u>us</u>.

3) A huge collision is <u>unlikely</u> in our lifetimes, but it's possible — there have certainly been big impacts in the past. Evidence for this includes:

   i)   <u>big craters</u>,

   ii) layers of <u>unusual elements</u> in rocks — these must have been 'imported' by an asteroid,

   iii) sudden changes in <u>fossil numbers</u> between adjacent layers of rock, as species suffer extinction.

4) <u>If</u> we get enough warning, we could try to <u>deflect</u> an NEO before it hits us. Scientists have various ideas about this — you could explode a nuclear bomb next to the object to 'nudge' it off course, or you could speed it up (or slow it down) so that it reaches Earth's orbit when we're out of the way.

Don't get asteroids, meteors and comets confused

It's serious business, this NEO stuff. In 2002, an asteroid narrowly missed us, and we only found it <u>12 days</u> beforehand — not very much time to put a plan together. Even if Bruce Willis had been on hand.

Magnetic Fields and Solar Flares

The Earth can protect us from a lot of nasties from space — e.g. cosmic rays.

The Earth is a bit like a big magnet

1) The Earth is surrounded by a <u>magnetic field</u> — a region where <u>magnetic materials</u> (like iron and steel) experience a <u>force</u>. Basically, this means the Earth acts like a big <u>bar magnet</u>.

2) Like all magnets, it has <u>north</u> and <u>south</u> <u>poles</u>.

> But... (concentrate now) the Earth's <u>south magnetic pole</u> is actually at the <u>North Pole</u>. This makes sense if you think about it... if you have a <u>compass</u> (or any other magnet), its north pole <u>points north</u> — because it's attracted towards a <u>south magnetic pole</u> (remember, opposites attract).

3) You can use a compass to tell the direction of the magnetic field. The needle points in the direction of the field.

4) The Earth <u>doesn't</u> actually have a <u>giant bar magnet</u> buried inside it. Its <u>core</u> contains a lot of <u>molten iron</u> and <u>nickel</u>, which move in <u>convection currents</u>. Scientists don't understand this fully, but they think that <u>electric currents</u> within this <u>liquid core</u> create the <u>magnetic field</u> (and vice versa).

There's more to the Earth than meets the eye

The Earth acting like a giant magnet sounds a bit funny at first — but it makes sense when you think about it. Remember, the Earth's south magnetic pole is at the North pole and vice versa. Just think of a compass — the needle's north pole points North, so it must be pointing at a south magnetic pole.

Magnetic Fields and Solar Flares

Earth's magnetic field **shields** us from **charged particles**

1) The surface of the Sun is a very unpleasant place — the Sun's constantly releasing enormous amounts of <u>energy</u> and <u>cosmic rays</u>. Cosmic rays are heavily <u>ionising</u>, and mostly consist of <u>charged particles</u>, though there's gamma rays and X-rays in there too.

2) The Earth's magnetic field does a good job of <u>shielding</u> us from a lot of the charged particles from the Sun by deflecting them away. But when charged particles in <u>cosmic rays</u> do hit the Earth's atmosphere they create <u>gamma rays</u>, forming part of the Earth's background radiation — see page 74.

DETLEV VAN RAVENSWAAY/
SCIENCE PHOTO LIBRARY

3) From time to time, massive <u>explosions</u> called <u>solar flares</u> also occur on the surface of the Sun. Solar flares release <u>vast</u> amounts of energy, some of it as <u>gamma rays</u> and <u>X-rays</u>.

4) Solar flares also give off <u>massive</u> clouds of <u>charged particles</u>. Some reach the Earth and produce <u>disturbances</u> in the Earth's <u>magnetic field</u>.

5) Solar flares can also damage <u>artificial satellites</u>, which we rely on for <u>all sorts</u> of things — for example... modern <u>communications</u>, <u>weather forecasting</u>, <u>spying</u>, <u>navigation</u> systems (such as GPS)...

6) The problem is that <u>electrons</u> and <u>ions</u> in solar flares can cause <u>surges</u> of <u>current</u> in a satellite's electrical circuitry. So satellites might need to be <u>shut down</u> to prevent damage during flares.

7) Solar flares also interact with the Earth's magnetic field, and can cause <u>power surges</u> in electricity distribution systems here on Earth (by means of <u>electromagnetic induction</u> — see page 40).

Cosmic rays cause the Aurora Borealis

1) Some charged particles in cosmic rays are <u>deflected</u> by the Earth's magnetic field and spiral down near the <u>magnetic poles</u>. Here, some of their <u>energy</u> is transferred to particles in the Earth's atmosphere, causing them to emit light — the <u>polar lights</u>.

2) The polar lights are shifting 'curtains' of light that appear in the sky. They're called the aurora borealis (<u>northern lights</u>) at the North Pole and the aurora australis at the South Pole. These displays are more dramatic during solar flares, when more cosmic rays arrive at the Earth.

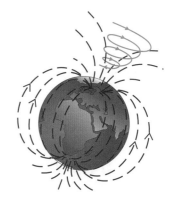

Solar flares can cause some big problems

In an old Scandinavian language, the word for 'northern lights' is 'herring flash' — people used to think the lights were <u>reflections</u> cast into the sky by large shoals of <u>herring</u>. That's not an <u>entirely</u> stupid idea, but <u>not</u> what you call the <u>accepted explanation</u> nowadays — so don't try it in the exam.

Beyond the Solar System

There's all sorts of exciting stuff in the Universe... Our whole Solar System is just part of a huge <u>galaxy</u>. And there are billions upon billions of galaxies. You should be realising now that the Universe is huge...

We're in the *Milky Way galaxy*

1) Our <u>Sun</u> is just one of <u>many billions</u> of <u>stars</u> which form the <u>Milky Way galaxy</u>. Our Sun is about halfway along one of the <u>spiral arms</u> of the Milky Way.

2) The <u>distance</u> between neighbouring stars in the galaxy is usually <u>millions of times greater</u> than the distance between <u>planets</u> in our Solar System.

3) The <u>force</u> which keeps the stars together in a galaxy is <u>gravity</u>, of course. And like most things in the Universe, galaxies <u>rotate</u> — a bit like a Catherine wheel.

The *whole universe* has *more than a billion galaxies*

1) Galaxies themselves are often <u>millions of times</u> further apart than the stars are within a galaxy.

2) So even the slowest among you will have worked out that the Universe is <u>mostly empty space</u> and is <u>really really BIG</u>.

Scientists measure distances in *space* using *light years*

1) Once you get outside our Solar System, the distances between stars and between galaxies are <u>so enormous</u> that kilometres seem too <u>pathetically small</u> for measuring them.

2) For example, the <u>closest</u> star to us (after the Sun) is about 40 000 000 000 000 kilometres away (give or take a few hundred billion kilometres). Numbers like that soon get out of hand.

3) So scientists use <u>light years</u> instead — a <u>light year</u> is the <u>distance</u> that <u>light travels</u> through a vacuum (like space) in one <u>year</u>. Simple as that.

4) If you work it out, 1 light year is equal to about 9 460 000 000 000 kilometres. Which means the closest star after the Sun is about <u>4.2 light years</u> away from us.

5) Just remember — a light year is a measure of <u>DISTANCE</u> (<u>not</u> time).

Stars can *explode* — and they sometimes leave *black holes*

1) When a <u>really big</u> star has used up all its fuel, it <u>explodes</u>. What's left after the explosion is really <u>dense</u> — sometimes so dense that <u>nothing</u> can escape its gravitational pull. It's now called a <u>black hole</u>.

See page 92 for more about the death of stars.

2) Black holes have a very <u>large mass</u> but their diameter is <u>tiny</u> in comparison.

3) They're <u>not visible</u> — even <u>light</u> can't escape their gravitational pull (that's why it's 'black').

4) Astronomers have to detect black holes in other ways — e.g. they can observe <u>X-rays</u> emitted by <u>hot gases</u> from other stars as they spiral into the black hole.

Warm-Up and Exam Questions

Here's some nice warm-up questions to get you into the swing of things before you try the exam questions.

Warm-Up Questions

1) Between which two planets in the Solar System are most asteroids found?
2) What are comets made of?
3) Give one difference between a comet and an asteroid.
4) What is a galaxy?
5) Why is a black hole black?

Exam Questions

1   The diagram shows a simple picture of part of the
     inner Solar System (not to scale).
     Match objects **1-4** with the descriptions given below.

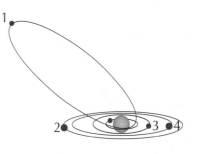

   A   Venus

   B   Earth

   C   Mars

   D   comet

(4 marks)

2   A light year is the

   A   size of a galaxy

   B   time taken for light to travel from the Sun to the Earth

   C   distance light travels in one year

   D   speed of the nearest galaxy to our own

(1 mark)

3   Solar flares occur on the surface of the Sun.

   (a)   What are solar flares?

(1 mark)

   (b)   (i)   What shields the Earth from most of the charged particles in solar flares?

(1 mark)

         (ii)   What effect can often be seen from near the North Pole as charged particles enter
               the atmosphere?

(1 mark)

   (c)   (i)   Give one use of artificial satellites.

(1 mark)

         (ii)   How can solar flares cause problems for artificial satellites?

(1 mark)

The Life Cycle of Stars

Stars go through <u>many traumatic stages</u> in their lives — just like teenagers.

Clouds of Dust and Gas

1) Stars <u>initially form</u> from clouds of <u>DUST AND GAS</u>.

Protostar

2) The <u>force of gravity</u> makes the gas and dust <u>spiral in together</u> to form a <u>protostar</u>. <u>Gravitational energy</u> has been converted into <u>heat energy</u>, so the <u>temperature rises</u>.

Main Sequence Star

3) When the <u>temperature</u> gets <u>high enough</u>, <u>hydrogen nuclei</u> undergo <u>nuclear fusion</u> to form <u>helium nuclei</u> and give out massive amounts of <u>heat and light</u>. A star is born. It immediately enters a <u>long stable period</u> where the <u>heat created</u> by the nuclear fusion provides an <u>outward pressure</u> to <u>balance</u> the <u>force of gravity</u> pulling everything <u>inwards</u>. In this stable period it's called a <u>MAIN SEQUENCE STAR</u> and it lasts <u>several billion years</u>. (The Sun is in the middle of this stable period — or to put it another way, the <u>Earth</u> has already had <u>half its innings</u> before the Sun <u>engulfs</u> it!)

Red Giant

4) Eventually the <u>hydrogen</u> begins to <u>run out</u> and the star then <u>swells</u> into a <u>RED GIANT</u>. It becomes <u>red</u> because the surface <u>cools</u>.

5) A <u>small star</u> like our Sun will then begin to <u>cool</u> and <u>contract</u> into a <u>WHITE DWARF</u> and then finally, as the <u>light fades completely</u>, it becomes a <u>BLACK DWARF</u>. (That's going to be really sad.)

Small stars

White Dwarf

Black Dwarf

Big stars

6) <u>Big stars</u>, however, start to <u>glow brightly again</u> as they undergo more <u>fusion</u> and <u>expand and contract several times</u>, forming <u>heavier elements</u> in various <u>nuclear reactions</u>. Eventually they'll <u>explode</u> in a <u>SUPERNOVA</u>.

new planetary nebula...   ...and a new solar system

Supernova

Neutron Star...

...or Black Hole

7) The <u>exploding supernova</u> throws the outer layers of <u>dust and gas</u> into space, leaving a <u>very dense core</u> called a <u>NEUTRON STAR</u>. If the star is <u>big enough</u> this will become a <u>BLACK HOLE</u> (see page 90).

8) The <u>dust and gas</u> thrown off by the supernova will form into <u>SECOND GENERATION STARS</u> like our Sun. The <u>heavier elements</u> are <u>only</u> made in the <u>final stages</u> of a <u>big star</u> just before and during the final <u>supernova</u>, so the <u>presence</u> of heavier elements in the <u>Sun</u> and the <u>inner planets</u> is <u>clear evidence</u> that our beautiful and wonderful world, with its warm sunsets and fresh morning dew, has all formed out of the snotty remains of a grisly old star's last dying sneeze.

What a star's like during its life determines what it becomes when it's dead

Erm. Now how do they know that exactly... Anyway, now you know what the future holds — our Sun is going to fizzle out, and it'll just get <u>very very cold</u> and <u>very very dark</u>. Great.

The Origins of the Universe

How did it all begin... Well, once upon a time, there was a really <u>Big Bang</u> — that's the <u>most convincing theory</u> we've got for how the Universe started.

The **Universe** seems to be **expanding**

As big as the Universe is, it looks like it's getting even bigger. All its galaxies seem to be moving away from each other. There's good evidence for this...

Light from other **galaxies** is **red-shifted**

1) When we look at <u>light from distant galaxies</u> we find that the <u>frequencies</u> are all <u>lower</u> than they should be — they're <u>shifted</u> towards the <u>red end</u> of the spectrum.

2) This is called the <u>red-shift</u>. (It's the same effect as a car's noise sounding <u>lower-pitched</u> when the car's <u>moving away</u> from you.)

3) <u>Measurements</u> of red-shift suggest that <u>all the galaxies</u> are <u>moving away from us</u> very quickly — and it's the <u>same result</u> whichever direction you look in.

More distant galaxies have **greater red-shifts**

1) <u>More distant</u> galaxies have <u>greater</u> red-shifts than nearer ones.

2) This means that more distant galaxies are <u>moving away faster</u> than nearer ones. All these findings indicate that the whole Universe is <u>expanding</u>.

There's **uniform microwave radiation** from all **directions**

This is another observation that scientists have made. It's not all that interesting in itself, but the theory that explains all this evidence definitely is.

1) Scientists have detected <u>low frequency electromagnetic radiation</u> coming from <u>all parts</u> of the Universe.

2) This radiation is in the <u>microwave</u> part of the EM spectrum. It's known as the <u>cosmic microwave background radiation</u>.

3) For complicated reasons this background radiation is <u>strong evidence</u> for an <u>initial Big Bang</u>.

4) As the Universe <u>expands and cools</u>, this background radiation 'cools' and <u>drops in frequency</u>.

There are microwaves in space — but not from ovens

I always thought the Universe was big enough already, but it's getting bigger all the time. How do we know? From clues like red-shift and microwave radiation — it'd be a bit difficult to go and measure.

The Origins of the Universe

Right now, all the galaxies are moving away from each other at great speed. But something must have got them going. That 'something' was probably a big explosion — so they called it the Big Bang...

It all **started off** with a very **big bang** (probably)

1) According to this theory, all the matter and energy in the Universe was compressed into a very small space. Then it exploded and started expanding.

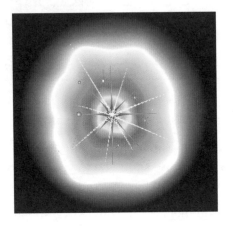

2) The expansion is still going on. We can use the current rate of expansion of the Universe to estimate its age. Our best guess is that the Big Bang happened about 14 billion years ago (though that might not be very accurate, as it's hard to tell how much the expansion has slowed down since the Big Bang).

3) The rate at which the expansion is slowing down is an important factor in deciding the future of the Universe. Without gravity the Universe would expand at the same rate forever. But as it is, all the masses in the Universe attract each other — and tend to slow the expansion down.

The **future** of the **Universe**...

1) The eventual fate of the Universe depends on how fast the galaxies are moving apart and how much total mass there is in it. But working this out is tricky — most of the mass appears to be invisible. Astronomers can only detect this dark matter by the way it affects the movement of things we can see.

2) Anyway, depending on how much mass there is, there are two ways the Universe could go:

Either a big crunch...

If there's enough mass compared to how fast the galaxies are currently moving, the Universe will eventually stop expanding — and begin contracting. This would end in a Big Crunch.

...or just cold lonely oblivion

If there's too little mass in the Universe to stop the expansion, it could expand forever with the Universe becoming more and more spread out into eternity.

Not everyone agrees but, the Big Bang seems to explain most evidence

Most scientists accept the idea of the Big Bang — it's the best way to explain the evidence we have at the moment. But if new evidence turns up, the theory could turn out to be rubbish. After all, there wasn't anyone around 14 billion years ago, taking photos and writing things down in a little notebook.

Warm-Up and Exam Questions

These warm-up questions are here to make sure you know the basics.
If there's anything you've forgotten, check up on the facts before you go any further.

Warm-Up Questions

1) What type of star will form a black hole?
2) What is the 'cosmic microwave background radiation'?
3) At the end of its main sequence stage, what does a star become?
4) What are the two possible fates for the Universe?
5) Our Sun is a second generation star. What does this mean?

Exam Questions

1   These sentences describe how a main sequence star is formed.
They are in the wrong order.

    1.   The gravitational energy is converted into heat energy.

    2.   The star is stable as the forces from pressure and gravity are balanced.

    3.   A large cloud of dust and gas collapses under gravity.

    4.   Nuclear fusion begins.

The correct order for these sentences is

A   3   4   2   1

B   1   2   3   4

C   1   4   2   3

D   3   1   4   2

(1 mark)

2   Viewed from Earth, other galaxies seem to be

A   moving away from us

B   moving towards us

C   contracting

D   expanding

(1 mark)

3   (a)   How do scientists believe the Universe started?

(1 mark)

(b)   Explain how the relationship between a galaxy's distance from Earth and its observed red shift suggests that the Universe is expanding.

(2 marks)

(c)   Scientists are unsure whether the Universe will expand forever or eventually begin to contract. Explain why.

(1 mark)

Exploring the Solar System

If you want to know what it's like on another planet, you have three options — peer at it from a distance (see page 98), get in a spaceship and go there yourself, or send a robot to have a peek...

Space exploration looks for **signs of life**

Part of the reason for exploring other planets it to try and find out if there's life out there.

1) <u>Sunlight</u> reflected from a planet, or refracted through its atmosphere, can give clues about what's on its <u>surface</u> or in its <u>atmosphere</u>.

2) Scientists look for <u>chemical changes</u> in a planet's atmosphere. Some changes in an atmosphere could be caused by things like volcanoes...

3) ...others are a <u>clue</u> that there might be life there (e.g. changing levels of <u>oxygen</u> and <u>carbon dioxide</u> in an atmosphere could be due to respiration or photosynthesis).

Little green men are the work of science-fiction — life on other planets is more likely to be simple organisms consisting of just a few cells.

Sometimes **manned spacecraft** are used...

1) The Solar System is <u>big</u> — so big that even radio waves (which travel at 300 000 000 m/s) take several <u>hours</u> to cross it. Even from <u>Mars</u>, radio signals take at least a couple of <u>minutes</u>.

2) But sending a <u>manned spacecraft</u> to Mars would take at least a couple of <u>years</u> (for a round trip).

3) The spacecraft would need to carry a lot of <u>fuel</u>, making it <u>heavy</u> — and <u>expensive</u>.

4) And it would be difficult keeping the astronauts <u>alive</u> and <u>healthy</u> for all that time.

... but keeping people **healthy** in space is **difficult**

Getting people into space is hard enough, but once they're up there, they need a lot of looking after...

1) The spacecraft would have to carry loads of <u>food</u>, <u>water</u> and <u>oxygen</u> (or be <u>very good</u> at <u>recycling</u>).

2) You'd need to regulate the <u>temperature</u> and remove <u>toxic gases</u> (e.g. CO_2) from the air.

3) The spacecraft would have to <u>shield</u> the astronauts from <u>cosmic rays</u> from the Sun.

4) Long periods in <u>low gravity</u> causes <u>muscle wastage</u> and loss of <u>bone tissue</u>.

5) Spending <u>ages</u> in a <u>tiny space</u>, with the <u>same people</u>, is psychologically <u>stressful</u>.

Space travel can be very stressful.

Exploring the Solar System

If sending people out to explore the solar system is too expensive (and downright tricky), a useful alternative is to send a robot to do the job for you.

Sending *unmanned probes* is much *easier*

Build a spacecraft. Pack as many instruments on board as will fit. Launch. That's the basic idea here.

1) 'Fly-by' missions are simplest — the probe passes close by an object but doesn't land. It can gather data on loads of things, including temperature, magnetic and gravitational fields and radiation levels.

2) Sometimes a probe is programmed to enter a planet's atmosphere. It might be designed to burn up after a while, having already sent back lots of data — about the atmosphere and radiation levels, say.

3) Some probes are designed to land on other planets (or moons, asteroids...). They often carry exploration rovers that can wander about, taking photos, etc. On Mars recently, NASA's two rovers (called 'Spirit' and 'Opportunity') were able to search for features of interest, e.g. evidence of water.

Advantages *of unmanned probes*

1) They don't have to carry food, water and oxygen (or people) — so more instruments can be fitted in.

2) They can withstand conditions that would be lethal to humans (e.g. extreme heat, cold or radiation levels).

3) They're cheaper — they carry less, they don't have to come back to Earth, and less is spent on safety.

4) If the probe does crash or burn up unexpectedly it's a bit embarrassing, and you've wasted lots of time and money, but at least no one gets hurt.

Disadvantages *of unmanned probes*

1) Unmanned probes can't think for themselves (whereas people are very good at overcoming simple problems that could be disastrous).

2) A spacecraft can't do maintenance and repairs — people can (as the astronauts on the Space Shuttle 'Discovery' had to do when its heat shield was damaged during takeoff).

Make sure you know the pros and cons of unmanned probes

When people first sent things into space, we began cautiously — in October 1957, Russia sent a small aluminium sphere (Sputnik 1) into orbit around the Earth. A month later, off went Sputnik 2, carrying the very first earthling to leave the planet — a small and unfortunate dog called Laika.

Looking into Space

There are various objects in space, and they emit or reflect different frequencies of EM radiation. And that's what you need to detect if you want to find out what's going on 'out there'.

Space telescopes have a clearer view than those on Earth

Telescopes help you to see distant objects clearly. But there can be problems...

1) If you're trying to detect light, Earth's atmosphere gets in the way — it absorbs a lot of the light coming from space before it can reach us.

2) Then there's pollution. Light pollution (light thrown up into the sky from streetlamps, etc.) makes it hard to pick out dim objects. And air pollution (e.g. dust particles) can reflect and absorb light coming from space. So to get the best view possible from Earth, a telescope should be on top of a mountain (where there's less atmosphere above it), and far away from any cities (e.g. on Hawaii).

MAGRATH PHOTOGRAPHY/
SCIENCE PHOTO LIBRARY

3) But to avoid the problem of the atmosphere, the thing to do is put your telescope in space, away from the mist and murk down here. The first space telescope (called Hubble) was launched by NASA in 1990. It can see objects that are about a billion times fainter than you can see unaided from Earth.

4) Hubble is an optical telescope (see the next page) and has a mirror. But because it gets a clear view into space, the mirror can be a lot smaller (and easier to make) than you'd need for a similar telescope on Earth.

5) It's not all plain sailing though. Getting a telescope safely into space is hard. And when things go wrong, it's difficult to get the repair men out. Hubble's first pictures were all fuzzy, because the mirror was the wrong shape. NASA had to send some astronauts up there to fix it.

6) So there are advantages to Earth-based telescopes — especially the fact that they're cheaper and easier to build and maintain.

Getting a telescope into space isn't easy

Most telescopes contain a lot of delicate, easily damaged parts. This is why it's so expensive to put them in space — they've got to be strong enough to withstand all the shaking on board the vehicle which takes them into orbit, but they need to be lightweight too.

Looking into Space

Putting a telescope in space isn't the only way to see past the atmosphere.
You can also try looking at different kinds of EM wave...

Different telescopes detect different types of *EM wave*

To get as full a picture of the Universe as possible, you need to detect different kinds of EM wave.

1) To see very <u>distant</u>, <u>faint</u> objects, you need a <u>very big telescope</u> or <u>lots of smaller ones</u> linked up — a bigger telescope 'collects' <u>more light</u> per second, making a brighter image.

2) Astronomers are usually keen to have images with good <u>resolution</u> (i.e. a lot of <u>detail</u>). The important thing here is how <u>big the telescope is</u> compared to the <u>wavelength</u> of the radiation. The <u>longer</u> the wave, the <u>bigger</u> the telescope you need to get the same level of detail.

Radio telescopes

<u>Radio telescopes</u> need to be <u>very large</u> — because radio waves are very <u>long</u>. But they're not too badly affected by the Earth's atmosphere, and they let astronomers observe objects which are too <u>faint</u> for optical telescopes (see below) to detect.

Optical telescopes

<u>Optical telescopes</u> detect <u>visible light</u>. They're used to look at objects close by and in other galaxies (though they are hampered by the Earth's atmosphere).

X-ray telescopes

<u>X-ray telescopes</u> are a good way to 'see' violent, <u>high-temperature events</u> in space, like <u>exploding stars</u>. But X-ray telescopes will <u>only</u> work <u>from space</u>, since the Earth's <u>atmosphere absorbs X-rays</u>. They're also <u>very expensive</u>, so there aren't many about.

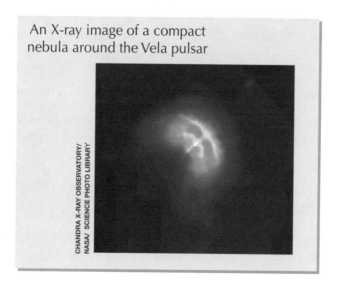

An X-ray image of a compact nebula around the Vela pulsar

CHANDRA X-RAY OBSERVATORY/ NASA/ SCIENCE PHOTO LIBRARY

Warm-Up and Exam Questions

It's that time again — a few easy questions to make sure you're awake, then some juicier, tougher ones. If you get stuck, go back to the right page and make sure you understand it.

Warm-Up Questions

1) Why are telescopes on Earth often built far away from towns and cities?
2) Why might an astronomer object to a campaign for brighter street lighting?
3) Give one reason why unmanned space probes are cheaper than manned space probes.
4) Describe one benefit of increasing the size of a telescope.
5) What does resolution mean in terms of the images produced by telescopes?

Exam Questions

1 Mars has been studied using unmanned probes since 1965 when Mariner 4 sent back images from its fly-by mission. However, no manned spacecraft has ever landed on Mars.

(a) If a manned spacecraft were to land on Mars, why would it be awkward for the astronauts on board to have a radio conversation with people on Earth?
(1 mark)

(b) Name two things that a spacecraft carrying astronauts to Mars would need to carry that an unmanned space craft would not need to carry.
(1 mark)

(c) Give two health effects that a person would suffer from spending long periods in low gravity.
(2 marks)

(d) Suggest one advantage of a manned space mission over an unmanned mission.
(1 mark)

2 The Hubble telescope was the first optical telescope to be launched into space.

(a) What is the main advantage of having an optical telescope in space rather than on Earth?
(1 mark)

(b) The Hubble space telescope was made as lightweight as possible. However, like all things in orbit, it is now weightless. Why was keeping its weight low important?
(1 mark)

3 The dish of the Arecibo radio telescope in Puerto Rico has a diameter of 305 m.

(a) It is not always practical to build as large telescopes as the Arecibo telescope. What alternative method is used to improve the resolution of Earth-based radio telescopes?
(1 mark)

(b) Explain in terms of wavelength why a radio telescope needs to be much larger than an optical telescope to achieve the same resolution.
(2 marks)

(c) Why would an Earth-based X-ray telescope not be useful?
(1 mark)

SECTION THREE — RADIOACTIVITY AND SPACE

Revision Summary for Section Three

Hopefully those little radiation facts will have penetrated your brain and stored themselves away — ready to be emitted at high speed in the exam. But there's only one way to find out. And you know what that is, I'll bet. This page isn't full of questions for nothing. Have fun...

1) Sketch an atom, labelling its protons, neutrons and electrons.

2) Explain what isotopes are. What does 'unstable' mean?

3) Oxygen atoms contain 8 protons. What is the difference between oxygen-16 and oxygen-18?

4) Radioactive decay is a totally random process. Explain what this means.

5) What is meant by ionisation?

6) Describe in detail the nature and properties of the three types of radiation: α, β and γ.

7) What substances could be used to block: a) α-radiation   b) β-radiation   c) γ-radiation?

8) List three sources of background radiation.

9) This data shows the count rate for a radioactive source at various times. Plot a graph of this data and use it to find the half-life of the substance.

Time (mins)	0	20	40	60	80	100	120
Count/minute	750	568	431	327	247	188	142

10) Explain how radiation damages the human body — a) at low doses, b) at high doses.

11) Describe what precautions you should take to protect yourself from alpha radiation, in a school laboratory.

12) State which types of radiation are used in each of the following, and explain why.
    a) medical tracers,   b) treating cancer,   c) detecting faults in aero engine turbine blades,
    d) sterilisation,   e) smoke detectors,   f) thickness control.

13)*An old bit of cloth was found to have 1 atom of C-14 to 80 000 000 atoms of C-12.
    If C-14 decays with a half-life of 5730 years, and the proportion of C-14 to C-12 in living material is 1 to 10 000 000, find the age of the cloth.

14) Describe the differences between stars and planets.

15) What and where are asteroids? What are comets made of? Sketch a diagram of a comet orbit.

16) What does NEO stand for? What are they and why do scientists lose sleep over them?

17) Sketch the magnetic field of the Earth. What do scientists think causes the Earth's magnetic field?

18) Explain how a solar flare can damage artificial satellites.

19) What's a light year?

20) How big is the Universe?

21) Why would a black hole form? Why's it called 'black'? How can you spot one?

22) Describe the first stages of a star's formation. Where does the initial energy come from?

23) What happens inside a star to make it so hot?

24) What is a 'main sequence' star? How long does it last? What happens after that?

25) What are the final two stages of a small star's life? What are the two final stages of a big star's life?

26) Why will our Sun never form a black hole?

27) What's the main theory of the origin of the Universe? Give two important bits of evidence for it.

28) What are the two possible futures for the Universe, and what do they depend on?

29) How could scientists investigate Neptune's atmosphere without actually sending someone there?

30) Explain 4 possible problems with going on a really long space journey.

31) Give two advantages and two disadvantages of manned space travel, compared to unmanned probes.

32) Why are space telescopes so expensive and sometimes so troublesome?

33) Describe one advantage and one disadvantage of optical, radio and X-ray telescopes.

* Answers on page 273.

Velocity and Acceleration

Speed and velocity aren't the same thing, you know. There's more to velocity than meets the eye.

Speed and velocity are both how fast you're going

Speed and velocity are both measured in m/s (or km/h or mph). They both simply say how fast you're going, but there's a subtle difference between them which you need to know:

> Speed is just how fast you're going (for example, 30 mph or 20 m/s), with no regard to the direction. But velocity must also have the direction specified, for example, 30 mph north or 20 m/s, 060°.

Seems a bit fussy I know, but they expect you to remember that distinction, so there you go.

Speed, distance and time — the formula:

$$\text{speed} = \frac{\text{distance}}{\text{time}}$$

You really ought to get pretty slick with this very easy formula. As usual the formula triangle version makes it all a bit of a breeze:

You just need to try and think up some interesting word for remembering the order of the letters in the triangle, s d t. Erm... sedit, perhaps... well, you think up your own.

Example

A cat skulks 20 m in 35 seconds. Find:

a) its speed.

b) how long it takes to skulk 75 m.

Answer: Using the formula triangle:

a) s = d ÷ t = 20 ÷ 35 = 0.57 m/s   b) t = d ÷ s = 75 ÷ 0.57 = 131 s = 2 min 11 s

A lot of the time we tend to use the words 'speed' and 'velocity' interchangeably.

For example, to calculate velocity you'd just use the above formula for speed instead.

Velocity and Acceleration

I expect you were anxiously wondering where the 'acceleration' bit had got to. Don't worry, here it is...

Acceleration is how quickly you're speeding up

Acceleration is definitely <u>not</u> the same as <u>velocity</u> or <u>speed</u>.
Every time you read or write the word <u>acceleration</u>, remind yourself:

"<u>Acceleration</u> is <u>completely different</u> from <u>velocity</u>. Acceleration is how <u>quickly</u>
the velocity is <u>changing</u>."

Velocity is a simple idea. Acceleration is altogether more <u>subtle</u>, which is why it's <u>confusing</u>.

Acceleration — the formula:

$$\text{acceleration} = \frac{\text{change in velocity}}{\text{time taken}}$$

Well, it's <u>just another formula</u>. Just like all the others.

Stick the three bits in a <u>formula triangle</u>, and hey presto:

Mind you, there are <u>two</u> tricky things with this one:

First there's the "ΔV", which means working out the
<u>'change in velocity'</u> (as shown in the example below),
rather than just putting a <u>simple value</u> for speed or
velocity in.

Secondly there's the <u>units</u> of acceleration, which are m/s².
<u>Not m/s</u>, which is <u>velocity</u>, but m/s². Got it? No? Let's try once more: <u>Not m/s</u>, <u>but m/s²</u>.

Example

A skulking cat accelerates steadily from 2 m/s to 6 m/s in 5.6 s. Find its acceleration.

<u>Answer:</u>   Using the formula triangle:

a = Δv ÷ t = (6 − 2) ÷ 5.6 = 4 ÷ 5.6 = <u>0.71 m/s²</u>

All pretty basic stuff, I'd say.

Remember, speed, velocity and acceleration are all different

<u>Speed cameras</u> measure the speed of motorists, using two photos taken a fraction of a second apart.
Lines painted on the road help them work out <u>how far</u> the car travelled between the photos — and so
<u>how fast</u> it was going. And the photos always have the car's number plate in them. Clever, eh?

D–T and V–T Graphs

Make sure you learn all these details about <u>distance-time</u> and <u>velocity-time graphs</u> really carefully. Make sure you can <u>distinguish</u> between the two, too.

Distance-time graphs

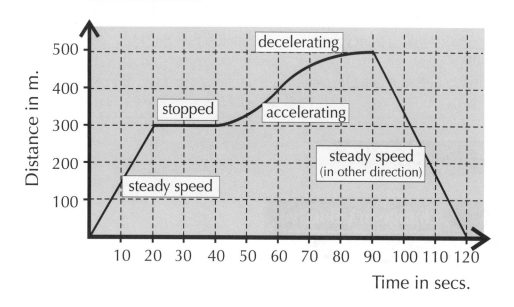

Very important notes:

1) <u>Gradient = speed</u>.

2) <u>Flat</u> sections are where it's <u>stopped</u>.

3) The <u>steeper</u> the graph, the <u>faster</u> it's going.

4) <u>Downhill</u> sections mean it's <u>going back</u> toward its starting point.

5) <u>Curves</u> represent <u>acceleration</u> or deceleration.

6) A <u>steepening</u> curve means it's <u>speeding up</u> (increasing gradient).

7) A <u>levelling off</u> curve means it's <u>slowing down</u> (decreasing gradient).

The **gradient** of a **distance-time** graph = the **speed**

For example, the <u>speed</u> of the <u>return</u> section of the graph is:

$$\underline{Speed} = \underline{gradient} = \frac{\text{vertical}}{\text{horizontal}} = \frac{500}{30} = \underline{16.7 \text{ m/s}}$$

Don't forget that you have to use the <u>scales</u> of the axes to work out the gradient.
<u>Don't</u> measure in <u>cm</u>!

D–T and V–T Graphs

Velocity-time graphs

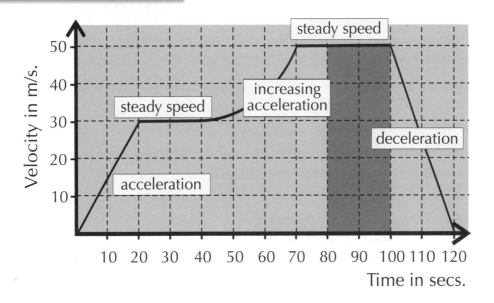

Very important notes:

1) <u>Gradient</u> = <u>acceleration</u>.

2) <u>Flat</u> sections represent <u>steady</u> speed.

3) The <u>steeper</u> the graph, the <u>greater</u> the <u>acceleration</u> or deceleration.

4) <u>Uphill</u> sections (/) are <u>acceleration</u> and <u>downhill</u> sections (\) are <u>deceleration</u>.

5) The <u>area</u> under any section of the graph (or all of it) is equal to the <u>distance</u> travelled in that <u>time</u> interval.

6) A <u>curve</u> means <u>changing acceleration</u>.

Calculating *acceleration*, *speed* and *distance* from a *V-T* graph

1) The <u>acceleration</u> represented by the <u>first section</u> of the graph is:

$$\text{Acceleration} = \text{gradient} = \frac{\text{vertical}}{\text{horizontal}} = \frac{30}{20} = 1.5 \text{ m/s}^2$$

2) The <u>speed</u> at any point is simply found by <u>reading the value</u> off the <u>velocity axis</u>.

3) The <u>distance travelled</u> in any time interval is equal to the <u>area</u> under the graph. For example, the distance travelled between $t = 80$ s and $t = 100$ s is equal to the <u>shaded area</u> which is equal to <u>1000 m</u>.

Don't get the two types and what they show confused

The tricky thing about these two types of graph is that they can look pretty much the <u>same</u> but represent <u>different kinds of motion</u>. Make sure you learn all the numbered points, and whenever you're reading a motion graph, check the <u>axis labels</u> carefully so you know which type of graph it is.

Mass, Weight and Gravity

The only thing that stops you flying off the planet into space is gravity. You'd be <u>very lost</u> without it.

Gravity is the force of attraction between all masses

<u>Gravity</u> attracts <u>all</u> masses, but you only notice it when one of the masses is <u>really really big</u>, e.g. a planet. Anything near a planet or star is <u>attracted</u> to it <u>very strongly</u>. This has <u>three</u> important effects:

1) On the surface of a planet, it makes all things <u>accelerate</u> towards the <u>ground</u> (all with the <u>same</u> acceleration, g, which is about <u>10 m/s^2 on Earth</u>).

2) It gives everything a <u>weight</u>.

3) It keeps <u>planets</u>, <u>moons</u> and <u>satellites</u> in their <u>orbits</u>. The orbit is a <u>balance</u> between the <u>forward</u> motion of the object and the force of gravity pulling it <u>inwards</u>.

Weight and mass are not the same

To understand this you must learn all these facts about mass and weight:

1) <u>Mass</u> is just the <u>amount of 'stuff'</u> in an object. For any given object this will have the same value <u>anywhere</u> in the Universe.

2) <u>Weight</u> is caused by the <u>pull</u> of gravity. In most questions the <u>weight</u> of an object is just the <u>force</u> of gravity pulling it towards the centre of the <u>Earth</u>.

3) An object has the <u>same</u> mass whether it's on <u>Earth</u> or on the <u>Moon</u> — but its <u>weight</u> will be <u>different</u>. A 1 kg mass will <u>weigh less</u> on the moon (about 1.6 N) than it does on <u>Earth</u> (about 10 N), simply because the <u>force</u> of gravity pulling on it is <u>less</u>.

4) Weight is a <u>force</u> measured in <u>newtons</u>. It's measured using a <u>spring balance</u> or <u>newton meter</u>. <u>Mass</u> is <u>not</u> a force. It's measured in <u>kilograms</u> with a <u>mass</u> balance (an old-fashioned pair of balancing scales).

The very important formula relating mass, weight and gravity

weight = mass × gravitational field strength

$$W = m \times g$$
(in <u>N</u>) (in <u>kg</u>) (in <u>N/kg</u>)

1) Remember, weight and mass are <u>not the same</u>. Mass is in <u>kg</u>, weight is in <u>newtons</u>.

2) The letter 'g' represents the <u>strength of gravity</u>, in N/kg — it's <u>different</u> for <u>different planets</u>. <u>On Earth</u> g ≈ 10 N/kg. <u>On the Moon</u>, where the gravity is weaker, g is only about 1.6 N/kg.

3) This formula is <u>hideously easy</u> to use:

<u>EXAMPLE:</u>   What is the weight, in newtons, of a 5 kg mass, both on Earth and on the Moon?

<u>ANSWER:</u>   'W = m × g'. So on Earth: W = 5 × 10 = <u>50 N</u>   (The weight of the 5 kg mass is 50 N.)

On the Moon: W = 5 × 1.6 = <u>8 N</u>   (The weight of the 5 kg mass is 8 N.)

See what I mean? Hideously easy — as long as you've learnt what all the letters mean.

Warm-Up and Exam Questions

Here's another set of questions to test your knowledge. Make sure you can answer them all before you go steaming on.

Warm-Up Questions

1) A boy hops 10 m in 5 seconds. What is his speed?
2) What are the units of acceleration? and of mass? and of weight?
3) What does the gradient of a distance-time graph show?
4) A car goes from 0 to 30 m/s in 6 seconds. Calculate its acceleration.
5) Name the force that keeps the Earth orbiting around the Sun.

Exam Questions

1 A racing car drives round a circular track of length 2400 m at a constant speed of 45 m/s.

(a) Explain why the car's velocity is not constant.

(1 mark)

(b) How long does the car take to travel once around the track?

(2 marks)

(c) On another lap, the car increases its speed from 45 m/s to 59 m/s over a period of 5 seconds. What is its acceleration?

(2 marks)

2 The graph below shows the distance of a shuttle-bus from its start point plotted against time.

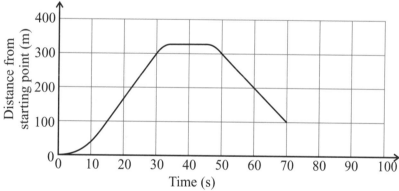

(a) Between 15 and 30 seconds:

  (i) how far does the bus travel?

(1 mark)

  (ii) how fast is the bus going?

(2 marks)

(b) For how long does the bus stop?

(1 mark)

(c) Describe the bus's speed and direction between 50 and 70 seconds.

(1 mark)

(d) Between 70 and 100 seconds, the bus slows, coming to a standstill at 100 s to finish up where it started. Show this on the graph.

(1 mark)

Exam Questions

3   The diagram below shows the velocity of a cyclist plotted against time.

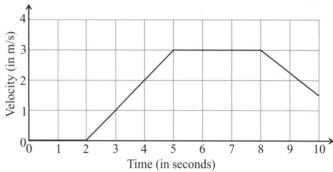

(a) Describe the motion of the cyclist between 5 and 8 seconds.

(1 mark)

(b) What is happening to the cyclist's speed between 8 and 10 seconds?

(1 mark)

(c) How far did the cyclist travel between 2 and 5 seconds.

(1 mark)

4   A spring increases in length when weights are suspended
from it, as shown. When a metal ball with a mass of 0.1 kg
is suspended from the spring, the spring stretches by 3 cm.

If the experiment was repeated on Mars, the spring would
only be stretched by 1.1 cm.

(a) Explain why the spring would stretch less on Mars,
given that the Earth's mass is 5.97×10^{24} kg and the mass of Mars is 6.42×10^{23} kg.

(3 marks)

(b) Estimate g on Mars, assuming that g on Earth is 10 m/s².

(2 marks)

5   A stone falls from the edge of a cliff. After falling for 1 second the stone has a downwards
velocity of 10 m/s.

(a) What is the stone's acceleration during the first second it falls?

(1 mark)

(b) Assuming no air resistance, what is the stone's velocity after three seconds of falling?

(1 mark)

(c) The stone has a mass of 0.12 kg. What is its weight?

(1 mark)

(d) What effect would doubling the stone's mass have on its acceleration due to gravity?

(1 mark)

6   Which of the following masses exert a gravitational attraction on other masses — the Sun,
the Earth, a human being, a feather, an atom?

(1 mark)

The Three Laws of Motion

Around about the time of the Great Plague in the 1660s, a chap called <u>Isaac Newton</u> worked out the <u>Three Laws of Motion</u>. At first they might seem kind of obscure or irrelevant, but to be perfectly blunt, if you can't understand these <u>three simple laws</u> then you'll never understand <u>forces and motion</u>.

First law — *balanced forces mean no change in velocity*

So long as the forces on an object are all <u>balanced</u>, then it'll just <u>stay still</u>, or else if it's already moving it'll just carry on at the <u>same velocity</u> — so long as the forces are all <u>balanced</u>.

1) When a train or car or bus or anything else is <u>moving</u> at a <u>constant velocity</u> then the <u>forces</u> on it must all be <u>balanced</u>.

2) Never let yourself entertain the <u>ridiculous idea</u> that things need a constant overall force to <u>keep</u> them moving — NO NO NO NO NO!

3) To keep going at a <u>steady speed</u>, there must be <u>zero resultant force</u> — and don't you forget it.

Second law — *a resultant force means acceleration*

If there is an <u>unbalanced force</u>, then the object will <u>accelerate</u> in that direction.

1) An <u>unbalanced</u> force will always produce <u>acceleration</u> (or deceleration).

2) This 'acceleration' can take <u>five</u> different forms: <u>Starting</u>, <u>stopping</u>, <u>speeding up</u>, <u>slowing down</u> and <u>changing direction</u>.

3) On a force diagram, the <u>arrows</u> will be <u>unequal</u>:

<u>Don't ever say:</u> "If something's moving there must be an overall resultant force acting on it".

Not so. If there's an <u>overall</u> force it will always <u>accelerate</u>.

You get <u>steady</u> speed from <u>balanced</u> forces.

The Three Laws of Motion

Three points which should be obvious:

1) The bigger the <u>force</u>, the <u>greater</u> the <u>acceleration</u> or <u>deceleration</u>.

2) The bigger the <u>mass</u>, the <u>smaller the acceleration</u>.

3) To get a <u>big</u> mass to accelerate <u>as fast</u> as a <u>small</u> mass it needs a <u>bigger</u> force. Just think about pushing <u>heavy</u> trolleys and it should all seem <u>fairly obvious</u>.

The overall unbalanced force is often called the resultant force

Any <u>resultant force</u> will produce <u>acceleration</u>, and this is the <u>formula</u> for it:

$$F = ma \quad \text{or} \quad a = F \div m$$

m = mass, a = acceleration, F is always the <u>resultant force</u>.

Resultant force is really important — especially for 'F = ma'

The notion of <u>resultant force</u> is a really important one for you to get your head round. It's not especially tricky, it's just that it seems to get <u>ignored</u>.

In most <u>real</u> situations there are at least <u>two forces</u> acting on an object along any direction. The <u>overall</u> effect of these forces will decide the <u>motion</u> of the object — whether it will <u>accelerate</u>, <u>decelerate</u> or stay at a <u>steady speed</u>. If the forces all point along the same direction, the '<u>overall effect</u>' is found by just <u>adding or subtracting</u> them. The overall force you get is called the <u>resultant force</u>. And when you use the <u>formula</u> '<u>F = ma</u>', F must always be the <u>resultant force</u>.

Example

A car of mass of 1750 kg has an engine which provides a driving force of 5200 N. At 70 mph the drag force acting on the car is 5150 N.

Find its acceleration:

a) when first setting off from rest   b) at 70 mph.

<u>Answer:</u> First draw a force diagram for both cases (no need to show the vertical forces):

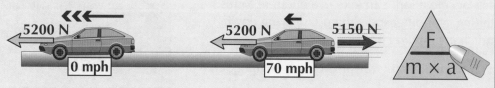

Work out the resultant force in each case, and apply 'F = ma' using the formula triangle:

a) Resultant force = 5200 N
   a = F/m = 5200 ÷ 1750 = <u>3.0 m/s²</u>

b) Resultant force = 5200 − 5150 = 50 N
   a = F/m = 50 ÷ 1750 = <u>0.03 m/s²</u>

The Three Laws of Motion

The third law — reaction forces

> If object A <u>exerts a force</u> on object B then object B exerts <u>the exact opposite force</u> on object A.

1) That means if you <u>push</u> something, say a shopping trolley, the trolley will <u>push back</u> against you, <u>just as hard</u>.

2) And as soon as you <u>stop</u> pushing, <u>so does the trolley</u>. Kinda clever really.

3) So far so good. The slightly tricky thing to get your head round is this — if the forces are always equal, <u>how does anything ever go anywhere</u>? The important thing to remember is that the two forces are acting on <u>different objects</u>.

Example

Think about a pair of ice skaters:

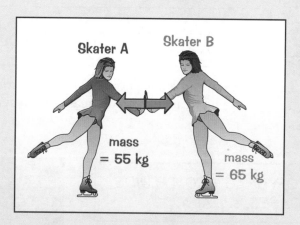

When skater A pushes on skater B (the '<u>action</u>' force), she feels an equal and opposite force from skater B's hand (the '<u>reaction</u>' force).

Both skaters feel the <u>same sized force</u>, in <u>opposite directions</u>, and so accelerate away from each other.

Skater A will be <u>accelerated</u> more than skater B, though, because she has a smaller mass — remember $F = ma$.

It's the same sort of thing when you go <u>swimming</u>. You <u>push</u> back against the <u>water</u> with your arms and legs, and the water pushes you forwards with an <u>equal-sized force</u> in the <u>opposite direction</u>.

Seems odd to think of walls pushing you and carts pulling you

This is like... proper Physics. It was <u>pretty fantastic</u> at the time — suddenly people understood how forces and motion worked and so they could work out the <u>orbits of planets</u> and everything. Inspired? No? Shame. Learn them anyway — you're really going to struggle in the exam if you don't.

Friction Forces

Friction is found nearly everywhere and it acts to <u>slow down</u> and <u>stop</u> moving objects. Sometimes friction is a pain, but at other times it's very helpful.

*Friction is always there to **slow things down***

1) If an object has <u>no force</u> propelling it along, it will always <u>slow down and stop</u> because of <u>friction</u> (unless you're out in space where there's no friction).

2) To travel at a <u>steady speed</u>, things always need a <u>driving force</u> to <u>counteract</u> the friction.

3) Friction acts in <u>three main ways</u>:

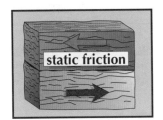

a) *FRICTION BETWEEN SOLID SURFACES WHICH ARE GRIPPING* (static friction)

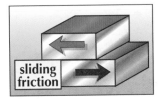

b) *FRICTION BETWEEN SOLID SURFACES WHICH ARE SLIDING PAST EACH OTHER*

You can <u>reduce</u> both these types of friction by putting a <u>lubricant</u> like <u>oil</u> or <u>grease</u> between the surfaces.

c) *RESISTANCE OR 'DRAG' FROM FLUIDS (LIQUIDS OR GASES, e.g. AIR)*

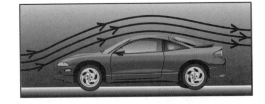

- The most important factor <u>by far</u> in <u>reducing drag in fluids</u> is keeping the shape of the object <u>streamlined</u>, like sports cars or boat hulls.

- Lorries and caravans have '<u>deflectors</u>' on them to make them more streamlined and reduce drag.

- <u>Roof boxes</u> on cars spoil their streamlined shape and so slow them down.

- For a given thrust, the <u>higher</u> the <u>drag</u>, the <u>lower</u> the <u>top speed</u> of the car.

- The <u>opposite extreme</u> to a sports car is a <u>parachute</u>, which is about as <u>high drag</u> as you can get — which is, of course, <u>the whole idea</u>.

In a <u>fluid</u>: <u>FRICTION ALWAYS INCREASES AS THE SPEED INCREASES</u> — and don't you forget it.

Friction's annoying when it's slowing down your boat, car or lorry...

... but it can be useful too. As well as stopping parachutists ending up as nasty messes on the floor, friction's good for <u>other stuff</u> — e.g. without it, you wouldn't be able to walk or run or skip or write.

Terminal Speed

Free-fallers reach a terminal speed

When free-falling objects <u>first set off</u> they have <u>much more</u> force <u>accelerating</u> them than <u>resistance</u> slowing them down. As the <u>speed</u> increases the resistance <u>builds up</u>.

This gradually <u>reduces</u> the <u>acceleration</u> until eventually the <u>resistance force</u> is <u>equal</u> to the <u>accelerating force</u> and then it won't be able to accelerate any more.
It will have reached its maximum speed or <u>terminal speed</u>.

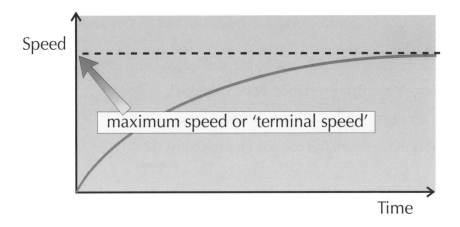

Speed

maximum speed or 'terminal speed'

Time

Terminal speed of falling objects depends on shape and area

1) The <u>accelerating force</u> acting on <u>all falling objects</u> is <u>gravity</u>, and it would make them all accelerate at the <u>same rate</u> if it wasn't for <u>air resistance</u>.

2) To prove this, on the Moon (where there's <u>no air</u>) hamsters and feathers dropped simultaneously will <u>hit the ground together</u>.

3) However, on Earth, <u>air resistance</u> causes things to fall at <u>different speeds</u>, and the <u>terminal speed</u> of any object is determined by its <u>drag</u> compared to its <u>weight</u>.

4) The drag depends on its <u>shape and area</u>.

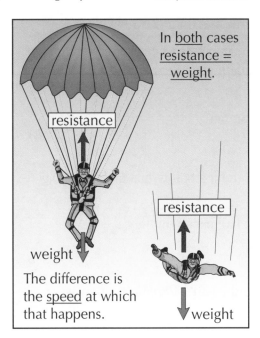

In <u>both</u> cases
<u>resistance =
weight</u>.

resistance

resistance

weight

The difference is
the <u>speed</u> at which
that happens.

weight

5) The best example is the <u>human skydiver</u>.

6) Without his parachute open he has quite a <u>small area</u> and a force equal to his <u>weight</u> pulling him down.

7) He reaches a <u>terminal speed</u> of about <u>120 mph</u>.

8) But with the parachute <u>open</u>, there's much more <u>air resistance</u> (at any given speed) and still only the same force pulling him down.

9) This means his <u>terminal speed</u> comes right down to about <u>15 mph</u>.

10) This is a <u>safe speed</u> to hit the ground at.

Warm-Up and Exam Questions

You've reached the halfway point of Section Four and it's time for some questions. Wahey.

Warm-Up Questions

1) What is the resultant force on a body moving at constant velocity?

2) What happens to the acceleration of a body if the resultant force on it is doubled?

3) A rowing boat is being pulled to shore by two people with a force of 3 N each.
   A force of 1 N is resisting the movement. What is the resultant force on the boat?

4) In which direction does friction act on a body — with or against the body's motion?

Exam Questions

1   Two parachutists, A and B, are members of the same club.

   (a) The diagram shows the forces acting on parachutist A.

   (i)   What is the resultant force acting on parachutist A?
         (1 mark)

   (ii) Is parachutist A moving? Explain your answer.
         (1 mark)

   (b) Parachutist B is in free fall. The total mass of parachutist B and equipment is 70 kg.

   (i)   What will the force of air resistance on parachutist B be when she reaches
         terminal velocity? Explain your answer.

                                                                                    (2 marks)

   (ii) Which parachutist, A or B, would have a higher terminal velocity?
         Explain your answer.

                                                                                    (3 marks)

   (c) Explain why a parachutist slows down when they open their parachute.

                                                                                    (1 mark)

2   Stefan weighs 600 newtons. He is accelerating upwards in a lift at 2.5 m/s^2.

   (a) The forces acting on Stefan are his weight and the upwards force exerted on him by
        the floor of the lift. Which force is greater? Explain your answer.

                                                                                    (1 mark)

   (b) What is the size of the resultant force acting on Stefan?

                                                                                    (2 marks)

3   Damien's cricket bat has a mass of 1.2 kg. He uses it to hit a ball with a mass of 160 g
     forwards with a force of 500 N.

   (a) Use Newton's third law to state the force that the ball exerts on the bat.
        Explain your answer.

                                                                                    (2 marks)

   (b) Which is greater — the acceleration of the bat or the ball? Explain your answer.

                                                                                    (2 marks)

Stopping Distances

If you <u>need to stop</u> in a <u>given distance</u>, then the <u>faster</u> you're going, the <u>bigger the braking force</u> you'll need. But real life's not quite that simple — there are loads of <u>other factors</u> too...

Many factors affect your total *stopping distance*

The stopping distance of a car is the distance covered in the time between the driver <u>first seeing</u> a hazard and the car coming to a <u>complete stop</u>. They're pretty keen on this for exams, so <u>learn it properly</u>.

The distance it takes to stop a car is divided into the <u>thinking distance</u> and the <u>braking distance</u>.

1) Thinking distance

"The distance the car travels in the split-second between a hazard appearing and the driver applying the brakes..."

It's affected by <u>three main factors</u>:

a) How <u>fast</u> you're going — obviously. Whatever your reaction time, the <u>faster</u> you're going, the <u>further</u> you'll go.

b) How <u>dopey</u> you are — This is affected by <u>tiredness</u>, <u>drugs</u>, <u>alcohol</u>, <u>old age</u>, and a <u>careless</u> attitude.

c) How the <u>visibility</u> is — lashing rain and oncoming lights, etc. make <u>hazards</u> harder to spot.

> The figures below for typical stopping distances are from the Highway Code. It's frightening to see just how long it takes to stop when you're going at 70 mph.

2) Braking distance

"The distance the car travels during its deceleration while the brakes are being applied..."

It's affected by <u>four main factors</u>:

a) How <u>fast</u> you're going — The <u>faster</u> you're going, the <u>further</u> you'll go before you stop. More details on page 124.

b) How <u>heavy</u> the load is — with the <u>same</u> brakes, <u>a heavily laden</u> vehicle takes <u>longer to stop</u>. A car won't stop as quickly if it's full of people and luggage and towing a caravan.

c) The <u>brakes'</u> condition — all brakes must be checked and maintained <u>regularly</u>. Worn or faulty brakes will let you down <u>catastrophically</u> just when you need them the <u>most</u>, i.e. in an <u>emergency</u>.

d) How good the <u>grip</u> is — this depends on <u>three things</u>:

- <u>road surface</u>, • <u>weather</u> conditions, • <u>tyres</u>.

So even at <u>30 mph</u>, you should drive no closer than <u>6 or 7 car lengths</u> away from the car in front — just in case. This is one reason why <u>speed limits</u> are so important, and some <u>residential areas</u> are now <u>20 mph zones</u>.

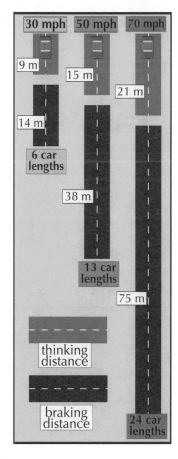

30 mph	50 mph	70 mph
9 m	15 m	21 m
14 m		
6 car lengths	38 m	75 m
	13 car lengths	24 car lengths

thinking distance

braking distance

Stop right there — and learn this page...

Leaves and oil spills on roads are <u>serious hazards</u> because they're <u>unexpected</u>. <u>Wet</u> or <u>icy roads</u> are always more <u>slippy</u> than dry roads, but often you only realise this when you try to <u>brake</u> hard. Tyres should have a minimum <u>tread depth</u> of <u>1.6 mm</u> — essential for getting rid of the <u>water</u> in wet conditions. Without <u>tread</u>, a tyre will simply <u>ride</u> on a <u>layer of water</u> and skid <u>very easily</u>.

Momentum and Collisions

A <u>large</u> rugby player running very <u>fast</u> is going to be a lot harder to stop than a scrawny one out for a Sunday afternoon stroll — that's <u>momentum</u> for you.

Momentum = mass × velocity

1) The <u>greater</u> the <u>mass</u> of an object and the <u>greater</u> its <u>velocity</u>, the <u>more momentum</u> the object has.

2) Momentum is a <u>vector</u> quantity — it has size <u>and</u> direction (like <u>velocity</u>, rather than speed).

Momentum (kg m/s) = mass (kg) × velocity (m/s)

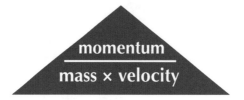

$$\frac{\text{momentum}}{\text{mass} \times \text{velocity}}$$

Momentum **before** = momentum **after**

<u>Momentum is conserved</u> when no external forces act, i.e. the total momentum <u>after</u> is the <u>same</u> as it was <u>before</u>.

Example 1:

Two skaters approach each other, collide and move off together as shown. At what velocity do they move after the collision?

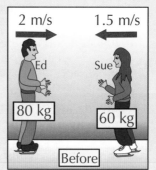

1) Choose which direction is <u>positive</u>. I'll say '<u>positive</u>' means '<u>to the right</u>'.

2) Total momentum <u>before</u> collision

= momentum of Ed + momentum of Sue

= (80 × 2) + [60 × (–1.5)] = <u>70 kg m/s</u>

3) Total momentum <u>after</u> collision

= momentum of Ed and Sue together

= <u>140 × v</u>

4) So 140 × v = 70,

i.e. <u>v = 0.5 m/s to the right</u>

Momentum and Collisions

You know what they say — when it comes to <u>momentum</u>, one example is never enough.

Example 2:

A gun fires a bullet as shown. At what speed does the gun move backwards?

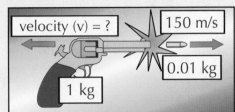

velocity (v) = ?   150 m/s   0.01 kg   1 kg

1) Choose which direction is <u>positive</u>.

   Again, I reckon '<u>positive</u>' means '<u>to the right</u>'.

2) <u>Total momentum before</u> firing

   = <u>0 kg m/s</u>

3) <u>Total momentum after</u> firing

   = momentum of bullet + momentum of gun

   = (0.01 × 150) + (1 × v)

   = <u>1.5 + v</u>

4) So 1.5 + v = 0, i.e. v = –1.5 m/s

   So the gun moves backwards at <u>1.5 m/s</u>.   *This is the gun's <u>recoil</u>.*

Forces cause changes in momentum

1) When a <u>force</u> acts on an object, it causes a <u>change</u> in momentum:

$$\text{Force acting (N)} = \frac{\text{change in momentum (kg m/s)}}{\text{time taken for change to happen (s)}}$$

2) A <u>larger</u> force means a <u>faster</u> change of momentum (and so a greater <u>acceleration</u>).

3) Likewise, if someone's momentum changes <u>very quickly</u> (like in a <u>car crash</u>), the <u>forces</u> on the body will be very <u>large</u>, and more likely to cause <u>injury</u>.

4) This is why cars are designed to slow people down over a <u>longer time</u> when they have a crash — the longer it takes for a change in <u>momentum</u>, the <u>smaller</u> the <u>force</u>.

| <u>CRUMPLE ZONES</u> crumple on impact, <u>increasing the time</u> taken for the car to stop. | <u>SEAT BELTS</u> stretch slightly, <u>increasing the time</u> taken for the wearer to stop. This <u>reduces the forces</u> acting on the chest. | <u>AIR BAGS</u> also slow you down more <u>gradually</u>. |

Momentum is a pretty fundamental bit of Physics — learn it well

Momentum is always <u>conserved</u> in collisions and explosions when there are no external forces acting. The bit at the bottom of the page is just another way of writing Newton's 2nd law of motion. Learn it.

Car Safety

A car speeding along a road has a lot of <u>kinetic energy</u>. If it suddenly drives into a wall, or a tree, or another car, that energy is used to push against the wall with an awful lot of <u>force</u>. And, as you saw earlier in the section, the wall <u>pushes back</u>...

Cars are designed to convert kinetic energy safely in a crash

1) If a car crashes, it will <u>slow down very quickly</u> — this means that a lot of <u>kinetic energy</u> is converted into other forms of energy in a <u>short amount of time</u>, which can be dangerous for the <u>people</u> inside.

2) Cars are <u>designed</u> to convert the <u>kinetic energy</u> of the car and its passengers in a way that is <u>safest</u> for the car's occupants.

3) <u>Crumple zones</u> are parts of the car at the front and back that crumple up in a <u>collision</u>. Some of the car's kinetic energy is converted into other forms of energy by the car body as it <u>changes shape</u>.

4) <u>Seat belts</u> and <u>air bags</u> slow the passengers down <u>safely</u> by converting their kinetic energy into other forms of energy over a longer period of time (see below). These safety features also <u>prevent</u> the passengers from hitting <u>hard surfaces</u> inside the car.

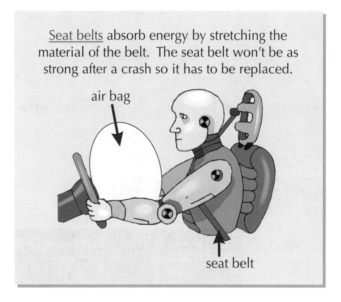

<u>Seat belts</u> absorb energy by stretching the material of the belt. The seat belt won't be as strong after a crash so it has to be replaced.

air bag

seat belt

Some safety features reduce forces

Cars have many <u>safety features</u> that are designed to <u>reduce the forces</u> acting on people involved in an accident. Smaller forces mean <u>less severe injuries</u>.

1) In a collision the <u>force</u> on the object can be lowered by <u>slowing it down</u> over a <u>longer time</u>.

2) <u>Safety features</u> in a car <u>increase the collision time</u> to <u>reduce the forces</u> on the passengers — e.g. crumple zones allow the car to slow down more <u>gradually</u> as parts of it change shape.

3) <u>Roads</u> can also be made safer by placing structures like <u>crash barriers</u> and <u>escape lanes</u> in dangerous locations (like on sharp bends or steep hills). These structures <u>increase the time</u> of any collision — which means the <u>collision force</u> is <u>reduced</u>.

Car Safety

Safety features come in two basic types — <u>active</u> and <u>passive</u>.
Both have helped to reduce the number of lives lost on Britain's roads.

Cars have *active* and *passive* safety features

1) Many cars have <u>active safety features</u> — these are features that <u>interact</u> with the way the car drives to help to <u>avoid a crash</u>, e.g. power assisted steering, traction control, etc.

2) <u>ABS brakes</u> are an active safety feature that <u>prevent skidding</u>. They help the driver to <u>stay in control</u> of the car when braking sharply. They can also give the car a <u>shorter stopping distance</u>.

3) A <u>passive safety feature</u> is any non-interactive feature of a car that helps to keep the <u>occupants</u> of the car <u>safe</u> — e.g. seat belts, air bags, headrests etc.

4) It's important that the <u>driver</u> of a car is <u>not distracted</u> when driving — and there are many features in a car that have been designed to keep the driver's attention firmly focused <u>on the road</u>.

5) Many new cars have <u>controls</u> on the <u>steering wheel</u> or on <u>control paddles</u> near the steering wheel. These features help drivers stay in control while using the stereo, electric windows, cruise control, etc.

steering wheel
stereo controls

driver's
airbag

There has been a recent *decrease* in UK car crash deaths

1) Safety features are <u>rigorously tested</u> to see how <u>effective</u> they are.

2) <u>Crash tests</u> have shown that wearing a <u>seat belt</u> reduces <u>fatalities</u> (deaths) by about 50%.

3) <u>Air bags</u> were shown to reduce the number of fatalities by about 30%.

4) There has been a <u>downward trend</u> in the number of deaths and serious injuries from <u>road traffic accidents</u> in the UK since 1980.

5) About <u>half as many people</u> are killed or seriously injured nowadays as in 1980 — this <u>reduction</u> is probably mainly due to the wide range of <u>safety features</u> found in cars today.

The UK is one of the best countries in Europe for car safety

The most important thing to learn here is that the <u>forces</u> acting on someone in a crash can be <u>reduced</u> by increasing the <u>collision time</u> — and there are loads of different safety features designed to do this...

Warm-Up and Exam Questions

1) What is meant by 'thinking distance' as part of the total stopping distance of a car?
2) What must be added to thinking distance to find the total stopping distance of a car?
3) Give the equation for momentum.
4) Name two safety features that increase the time taken for the car driver to stop in a collision.
5) Give an example of an active safety feature in a car.

Exam Questions

1   The graph below shows how thinking distance and stopping distance vary with speed.

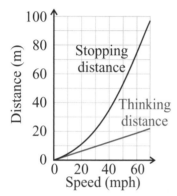

(a) Use the graph to determine the following distances for a car travelling at 40 mph.

  (i)   Thinking distance

  (ii) Total stopping distance

  (iii) Braking distance

(3 marks)

(b) Which is greater at 50 miles per hour, thinking distance or braking distance?

(1 mark)

(c) Is stopping distance proportional to speed? How can this be seen from the graph?

(2 marks)

2   Two ice hockey players are skating towards the puck. Player A has a mass of 100 kg and is travelling right at 6 m/s. Player B has a mass 80 kg and is travelling left at 9 m/s.

(a) Calculate the momentum of:

  (i)   Player A

(1 mark)

  (ii) Player B

(1 mark)

(b) The two players collide and become joined together.

  (i)   What is the speed of the two joined players just after the collision?

(2 marks)

  (ii) Which direction do they move in?

(1 mark)

3   (a) Why are seat belts made of a slightly stretchy material?

(1 mark)

(b) What is the difference between active and passive safety features?

(2 marks)

Work and Potential Energy

Work (like a lot of things) means something slightly <u>different</u> in Physics than it does in everyday life.

Doing **work** involves **transferring energy**

When a <u>force</u> moves an <u>object</u>, <u>ENERGY IS TRANSFERRED</u> and <u>WORK IS DONE</u>.

That statement sounds far more complicated than it needs to. Try this:

1) Whenever something <u>moves</u>, something else is providing some sort of '<u>effort</u>' to move it.

2) The thing putting the <u>effort</u> in needs a <u>supply</u> of energy (like <u>fuel</u> or <u>food</u> or <u>electricity</u>, etc.).

3) It then does '<u>work</u>' by <u>moving</u> the object — and one way or another it <u>transfers</u> the energy it receives (as fuel) into <u>other forms</u>.

4) Whether this energy is transferred '<u>usefully</u>' (e.g. by <u>lifting a load</u>) or is '<u>wasted</u>' (e.g. lost as <u>heat</u> through <u>friction</u>), you can still say that '<u>work is done</u>'. Just like Batman and Bruce Wayne, '<u>work done</u>' and '<u>energy transferred</u>' are indeed '<u>one and the same</u>'.   (And they're both given in <u>joules</u>.)

It's just **another trivial formula**:

work done = force × distance

$$\frac{W}{F \times d}$$

Whether the force is <u>friction</u> or <u>weight</u> or <u>tension in a rope</u>, it's always the same. To find how much <u>energy</u> has been <u>transferred</u> (in J), just multiply the <u>force in N</u> by the <u>distance moved in m</u>.

Example

Some naughty kids drag an old tractor tyre 5 m over rough ground. They pull with a total force of 340 N. Find the energy transferred.

<u>Answer</u>: W = F × d = 340 × 5 = <u>1700 J</u>.

Work and Potential Energy

Potential energy is the energy something gets when it is lifted. Think about it —
if you raise something to a great height and then let go... suddenly it has a lot of energy.

Potential energy is energy due to height

potential energy = mass × gravitational field strength × height

$$\frac{\text{P.E.}}{m \times g \times h}$$

The proper name for this kind of 'potential energy' is gravitational potential energy
(as opposed to 'elastic potential energy' or 'chemical potential energy', etc.).

On Earth, the gravitational field strength, g, is approximately 10 N/kg.

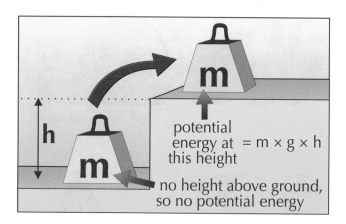

potential
energy at $= m \times g \times h$
this height

no height above ground,
so no potential energy

Example

A sheep of mass 47 kg is slowly raised through 6.3 m.
Find the gain in potential energy.

Answer: Just plug the numbers into the formula:

P.E. = mgh = 47 × 10 × 6.3 = 2961 J
(joules because it's energy)

Revising is a type of work too — and it certainly uses up energy

Remember 'energy transferred' and 'work done' are the same thing. If you need a force to make
something speed up (p.109), all that means is that you need to give it a bit of energy. Makes sense.

Kinetic Energy

<u>Kinetic energy</u> is the energy something has when it is moving. Think about it — if you bump into something <u>heavy</u> that's rushing downhill at <u>top speed</u>, it will do a lot more damage than if you bump into something <u>light</u> that's just <u>sitting there</u>.

Kinetic energy is energy of movement

Anything that's <u>moving</u> has <u>kinetic energy</u>.
There's a slightly <u>tricky formula</u> for it, so you have to concentrate a little bit <u>harder</u> for this one.
But hey, that's life — it can be tough sometimes:

$$\textbf{kinetic energy} = \tfrac{1}{2} \times \textbf{mass} \times \textbf{speed}^2$$

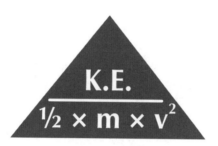

Example

A car with a mass of 2450 kg is travelling at 38 m/s.
Calculate its kinetic energy.

38 m/s

2450 kg

<u>Answer</u>: It's easy. You just plug the numbers into the formula.
Watch the 'v²' though...

K.E. = ½mv²

= ½ × 2450 × 38²

= <u>1 768 900 J</u>   (joules because it's <u>energy</u>)

Remember, the <u>kinetic energy</u> of something depends both on <u>mass</u> and <u>speed</u>.
The <u>more it weighs</u> and the <u>faster it's going</u>, the <u>bigger</u> its kinetic energy will be.

small mass, not fast
low kinetic energy

big fast lorries Ltd

big mass, very fast
high kinetic energy

Kinetic Energy

Stopping distances *increase alarmingly* with *extra speed*
— mainly because of the *v²* bit in *K.E.* = ½*mv²*

To stop a car, the <u>kinetic energy</u>, ½mv², has to be <u>converted to heat energy</u> at the <u>brakes and tyres</u>:

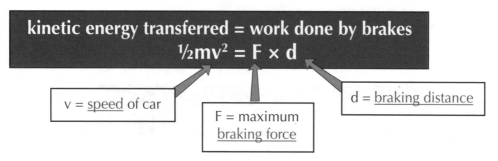

kinetic energy transferred = work done by brakes
$$\tfrac{1}{2}mv^2 = F \times d$$

v = <u>speed</u> of car

F = maximum <u>braking force</u>

d = <u>braking distance</u>

<u>Learn this really well</u>:

1) If you <u>double the speed</u>, you double the value of v, but the v² means that the <u>kinetic energy</u> is actually increased by a factor of <u>four</u>.

2) However, 'F' is always the <u>maximum possible</u> braking force which <u>can't</u> be increased, so <u>d</u> must also increase by a factor of <u>four</u> to make the equation <u>balance</u>.

3) So, if you go <u>twice as fast</u>, the <u>braking distance</u> 'd' must increase by a <u>factor of four</u> to dissipate the extra <u>kinetic energy</u>.

Falling objects convert *P.E.* into *K.E.*

When something falls, its <u>potential energy</u> is <u>converted</u> into <u>kinetic energy</u>. So the <u>further</u> it falls, the <u>faster</u> it goes.

In practice, some of the P.E. will be <u>dissipated</u> as <u>heat</u> due to <u>air resistance</u>, but in exam questions they'll likely say you can <u>ignore</u> air resistance, in which case you'll just need to remember this <u>simple</u> and <u>really quite obvious</u> formula:

kinetic energy <u>gained</u> = potential energy <u>lost</u>

K.E. and P.E. are closely linked in falling objects
This page also explains why braking distance goes up so much with speed. Bet you've been dying to find that out — and now you know. What you probably <u>don't</u> know yet, though, is that rather lovely formula at the top of the last page. I mean, it's got more than three letters in it — how special is that?

Roller Coasters

Roller coasters transfer energy

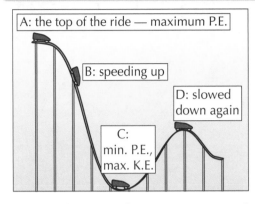

A: the top of the ride — maximum P.E.

B: speeding up

D: slowed down again

C: min. P.E., max. K.E.

1) At the top of a roller coaster (position A) the carriage has lots of <u>gravitational potential energy</u> (P.E.).

2) As the carriage descends to position B, P.E. is transferred to <u>kinetic energy</u> (K.E.) and the carriage speeds up.

3) Between positions B and C the carriage keeps <u>accelerating</u> as its P.E. is converted into K.E.

4) If you <u>ignore</u> any <u>air resistance</u> or <u>friction</u> between the carriage and the track, then all the P.E. the carriage had at A will be converted to K.E. by the time it reaches C.

5) A real carriage loses some energy through friction on the way down the hill. It needs to have enough <u>kinetic energy</u> left at point C to carry it up the hill again to D.

You can calculate the maximum speed of a roller coaster

Example

Look at the diagram of the roller coaster. Let's say that the vertical drop between positions A and C is 20 m. If a carriage of mass 500 kg starts from rest at A, what is the maximum possible speed of the carriage at position C? (Take g = 10 N/kg.)

Answer: First things first. The question asks for the <u>maximum possible speed</u> — that's when there's no friction, so you know potential energy at A = kinetic energy at C.

There are four key steps to this method — and you need to learn them:

Step 1 Find the P.E. lost: P.E. = mgh = 500 × 10 × 20 = <u>100 000 J</u>. This must also be the K.E. gained.

Step 2 Equate the number of joules of K.E. gained to the K.E. formula, '$\frac{1}{2}mv^2$':
$100\,000 = \frac{1}{2}mv^2$

Step 3 Stick the numbers in:
$100\,000 = \frac{1}{2} \times 500 \times v^2$ or $100\,000 = 250 \times v^2$

$100\,000 \div 250 = v^2$ so $v^2 = 400$

Step 4 Square root: $v = \sqrt{400}$ = <u>20 m/s</u>

Designers have to think about safety and the local community

When designing rides it's important to consider:

(1) <u>SAFETY FEATURES</u>: These can be built into the rides themselves, e.g. <u>headrests</u> and <u>seat belts</u>. Or there might be safety restrictions on the people who can go on a ride, e.g. <u>height restrictions</u>.

(2) <u>SOCIAL EFFECTS</u>: On the positive side, a new theme park will provide <u>jobs for locals</u> and <u>entertainment</u> for the area. It will <u>attract tourists</u> who will spend money in local businesses. On the downside, there will be an increase in <u>traffic</u>, <u>noise</u> and <u>litter</u>.

Warm-Up and Exam Questions

It's question time again. A few short ones to get you started, then some tougher ones to make sure you've really learnt your stuff.

Warm-Up Questions

1) Why can 'work done' be measured in the same units as energy?
2) What is gravitational potential energy?
3) Give the formula for gravitational potential energy.
4) If the speed of a car triples, what happens to its kinetic energy?
5) Why is the speed of a roller coaster greatest at the lowest point of the track?

Exam Questions

1   A train with a mass of 40 000 kg drives 700 m while accelerating at 1.05 m/s^2.

(a) Calculate the driving force acting on the train. Ignore any friction.

(2 marks)

(b) Find the work done by the driving force.

(1 mark)

(c) The train reaches a constant speed at which its kinetic energy is 29 400 000 J. It then decelerates with a braking force of 29 400 N. Calculate its braking distance.

(2 marks)

(d) How would the braking distance change if the train was only going half as fast when it began to decelerate?

(1 mark)

(e) What form of energy is the train's kinetic energy transformed into when the brakes are applied?

(1 mark)

2   A boy with a mass of 45 kg climbs up a 12 m high flight of steps.

(a) How much potential energy has the boy gained?

(2 marks)

(b) How much work has he done against gravity?

(1 mark)

3   A roller coaster train drives from ground level to the highest point of the ride, 26.45 m above the ground. The train pauses at this point, where it has 529 000 J of potential energy.

(a) What is the mass of the train?

(2 marks)

(b) The train rolls down the track until it reaches ground level.

(i)   What is its kinetic energy at ground level? (Ignore friction and air resistance.)

(1 mark)

(ii) Find the train's speed at ground level.

(2 marks)

Power

Force, energy, work, momentum, acceleration — your head must be spinning.
So I'll help you out by adding <u>power</u> into the mix.

Power *is the 'rate of doing work' — i.e. how much per second*

<u>POWER</u> is <u>not</u> the same thing as <u>force</u> or <u>energy</u>. A <u>powerful</u> machine is not necessarily one which can exert a strong <u>force</u> (though it does usually end up that way).

A <u>POWERFUL</u> machine is one which transfers <u>a lot of energy in a short space of time</u>.

This is the <u>very easy formula</u> for power:

$$\text{power} = \frac{\text{work done}}{\text{time taken}}$$

Example

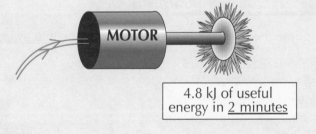

A motor transfers 4.8 kJ of useful energy in 2 minutes. Find its power output.

<u>Answer</u>: P = W ÷ t

= 4800 ÷ 120

= <u>40 W</u> (or 40 J/s)

4.8 kJ of useful energy in <u>2 minutes</u>

(Note that the kJ had to be turned into J, and the minutes into seconds.)

Power *is measured in watts (or J/s)*

1) The proper unit of power is the <u>watt</u>.

2) <u>One watt = 1 joule of energy transferred per second</u>.

3) <u>Power</u> means 'how much energy <u>per second</u>', so <u>watts</u> are the same as '<u>joules per second</u>' (J/s).

4) Don't ever say 'watts per second' — that would be <u>nonsense</u>.

Power

Calculating your **power output**

Both cases use the same formula:

$$\text{POWER} = \frac{\text{ENERGY TRANSFERRED}}{\text{TIME TAKEN}} \quad \text{or} \quad P = \frac{E}{t}$$

a)   The timed run **upstairs**:

In this case the 'energy transferred' is simply the potential energy you gain (= mgh).

Hence, power = mgh/t

Power output
= energy transferred / time
= mgh / t
= $(62 \times 10 \times 12) \div 14$
= 531 W

time taken = 14 s

62 kg

12 m

b)   The timed **acceleration**:

This time the energy transferred is the kinetic energy you gain (= ½mv²).

Hence, power = ½mv²/ t

Power output
= energy transferred / time
= ½mv² / t
= $(\frac{1}{2} \times 62 \times 8^2) \div 4$
= 496 W

62 kg

0 ➡ 8m/s

time taken = 4 s

Cars have different **power ratings**

1) The size and design of car engines determine how powerful they are.

2) The more powerful an engine, the more energy it transfers from its fuel every second, so (usually) the higher the fuel consumption (see next page).

3) E.g. the power output of a typical small car will be around 50 kW and a sports car will be about 100 kW (some are much higher).

Watts should be familiar to you thanks to the humble light bulb

The power of a car isn't always measured in watts — sometimes you'll see it in a funny unit called brake horsepower. James Watt defined 1 horsepower as the work done when a horse raises a mass of 550 lb (250 kg) through a height of 1 ft (0.3 m) in 1 second... as you do. I'd stick to watts if I were you.

Fuels for Cars

Cars and lorries would be pretty <u>useless</u> if they didn't have any <u>fuel</u> to get them moving...

Most cars run on *fossil fuels*

1) <u>All vehicles</u> need a <u>fuel</u> to make them move — most cars and lorries use <u>petrol</u> or <u>diesel</u>.

2) Petrol and diesel are fuels that are <u>made from oil</u>, which is a <u>fossil fuel</u>. The <u>pollution</u> released when these fuels are <u>burnt</u> can cause <u>environmental problems</u> like acid rain and climate change.

3) Fossil fuels are <u>non-renewable</u>, so one day they'll <u>run out</u> — not good if your car runs on petrol.

4) To avoid these <u>problems</u>, scientists are developing engines that run on <u>alternative types of fuel</u>, such as <u>alcohol</u>, <u>liquid petroleum gas</u> (LPG), <u>hydrogen</u> and '<u>biodiesel</u>'. They're not perfect, though. <u>LPG</u> still comes from fossil fuels — it just has lower emissions. <u>Hydrogen</u> can be produced by the electrolysis of a very dilute acid, but that takes energy that's likely to come from fossil fuels.

5) A few vehicles use <u>large batteries</u> to power <u>electric motors</u>. These vehicles don't release any <u>pollution</u> when they're driven, but their <u>batteries</u> need to be <u>charged</u> using electricity. This electricity is likely to come from <u>power stations</u> that do pollute.

Fuel consumption is all about the amount of *fuel used...*

1) The <u>fuel consumption</u> of a car is usually defined as the <u>distance travelled</u> using a <u>certain amount of fuel</u>. Fuel consumption is often given in <u>miles per gallon</u> (mpg) or <u>litres per 100 km</u> (l/100 km) — e.g. a car with a fuel consumption of 5 l/100 km will travel 100 km on 5 litres of fuel.

2) A car's fuel consumption <u>depends</u> on <u>many different things</u> — e.g. the type and size of the engine, how the car is driven, the shape and weight of the car, etc.

3) A car will have a <u>high fuel consumption</u> if a <u>large force</u> is needed to move it. Using $F = ma$, you can see that the force needed to move a car depends on the <u>mass</u> of the car and its <u>acceleration</u>.

4) A <u>heavy car</u> will need a greater <u>force</u> to accelerate it by a given amount than a lighter car, so the <u>fuel consumption</u> will be higher for the heavy car.

5) <u>Driving style</u> also affects the fuel consumption — big accelerations need more force and so use more fuel. <u>Frequent braking and acceleration</u> (e.g. when driving in a town) will <u>increase</u> fuel consumption.

6) <u>Opening the windows</u> will increase a car's fuel consumption — this is because <u>more energy</u> will be needed to overcome the increase in <u>air resistance</u>.

7) The <u>speed</u> a car's travelling at affects fuel consumption as well. Cars work <u>more efficiently</u> at some speeds than others — the most efficient speed is usually between 40 and 55 mph.

Examples of how <u>car design</u> can improve fuel consumption:

The <u>shape</u> of modern cars reduces air resistance.

<u>Engines</u> with better fuel efficiency are constantly being designed.

Cars are made from lightweight <u>materials</u>.

Lower fuel consumption is cheaper and better for the environment

You might get asked how to reduce the fuel consumption of a car, so it's important that you remember the different ways that fuel consumption can be affected — e.g. <u>friction</u>, <u>air resistance</u>, <u>weight</u>, etc.

Warm-Up and Exam Questions

It's very nearly the end of Section Four. But don't shed a tear — try these questions instead.

Warm-Up Questions

1) What is measured in watts?
2) Why does frequent braking and acceleration in a vehicle lead to higher fuel consumption?
3) Why are scientists searching for alternatives to petrol and diesel as fuel?
4) State two different units that are used to describe fuel consumption.
5) How do battery powered cars cause pollution indirectly?

Exam Questions

1   A hotel service lift raises a sack of laundry of mass 15 kg through 2 m in 12 seconds.

   (a)   How much work is done on the laundry?

   (2 marks)

   (b)   What is the power output of the lift?

   (2 marks)

   (c)   The service lift is powered by an electric motor.
         Describe the energy changes that occur when the sack is raised.

   (2 marks)

2   A girl is exercising in the park. Her mass is 52 kg.
    Calculate her power output during the following activities:

   (a)   Running from rest to 7 m/s in 3.5 seconds.

   (2 marks)

   (b)   A press-up where she lifts her centre of mass by 0.2 m in 0.7 seconds.

   (2 marks)

   (c)   Running up a 7 m high hill in 45 s.

   (2 marks)

3   The Marston family drive 240 km from their house to the seaside every year.

   (a)   One year they used 15.5 litres of fuel on the journey to the seaside.
         Calculate their fuel consumption.

   (1 mark)

   (b)   The next year they had a loaded roof rack on top of their car to carry extra luggage.
         Give two reasons why the roof rack would have increased their fuel consumption.

   (2 marks)

   (c)   Mr Marston calculates that their car transfers 4800 kJ of energy from its fuel every
         minute. What is the power output of the car in kilowatts?

   (2 marks)

Revision Summary for Section Four

Yay — revision summary! I <u>know</u> these are your favourite bits of the book, all those lovely questions. There are lots of formulas and laws and picky little details to learn in this section. So, practise these questions till you can do them all standing on one leg with your arms behind your back while being tickled on the nose with a purple ostrich feather. Or something.

1) Write down the formula for working out speed. What's the difference between speed and velocity?

2)* a) Find the speed of a partly chewed mouse which hobbles 3.2 metres in 35 seconds.
    b) Find how far the mouse would go in 25 minutes.

3)* A speed camera is set up in a 30 mph (13.3 m/s) zone. It takes two photographs 0.5 s apart. A car travels 6.3 m between the two photographs. Was the car breaking the speed limit?

4) What is acceleration? What's the unit for it?

5) a) Write down the formula for acceleration.
   b)* What's the acceleration of a soggy pea flicked from rest to a speed of 14 m/s in 0.4 seconds?

6) What does the area under a velocity-time graph tell you?

7) Sketch a typical velocity-time graph and point out all the important points.

8) Explain how to find acceleration from a velocity-time graph.

9) What is gravity? List three effects gravity produces.

10) Explain the difference between mass and weight. What units are they measured in?

11) What's the formula for weight? Illustrate it with a worked example of your own.

12) If an object has zero resultant force on it, can it be moving? Can it be accelerating?

13) a) Write down the formula relating resultant force and acceleration.
    b)* A force of 30 N pushes a trolley of mass 4 kg. What will be its acceleration?

14) Explain what a reaction force is.

15) What could you do to reduce the friction between two surfaces?

16) Describe how air resistance is affected by speed.

17) What is 'terminal speed'? What two main factors determine the terminal speed of a falling object?

18) What are the two different parts of the overall stopping distance of a car?

19) List three factors that affect each of the two parts of the stopping distance of a car.

20)* Write down the formula for momentum. Find the momentum of a 78 kg sheep moving at 23 m/s.

21) If the total momentum of a system before a collision is zero, what is the total momentum of the system after the collision?

22) Explain how air bags, seat belts and crumple zones reduce the risk of serious injury in a car crash.

23)* A crazy dog dragged a big branch 12 m over the next-door neighbour's front lawn, pulling with a force of 535 N. How much work was done on the branch?

24)* Calculate the increase in potential energy when a box of mass 12 kg is raised through 4.5 m. (Assume g = 10 N/kg.)

25) How does the kinetic energy formula explain the effect of speed on the braking distance of a car?

26)* At the top of a roller coaster ride, a stationary carriage has 150 kJ of gravitational potential energy (P.E.). Ignoring friction, how much kinetic energy must it have at the bottom (when P.E. = 0)?

27)* A 600 kg roller coaster carriage is travelling at 40 m/s. What is the maximum height it could climb if all its kinetic energy is transferred to gravitational potential energy?

28) State two safety features associated with theme park rides.

29)* An electric motor uses 540 kJ of electrical energy in 4.5 minutes. What is its power consumption?

30)* Calculate the power output of a 78 kg sheep that runs 20 m up a staircase in 16.5 seconds.

31) Electric vehicles don't give out polluting gases directly, but they still cause pollution. Explain why.

32) Give three factors that affect the fuel consumption of a car.

* Answers on page 274.

Static Electricity

Static electricity is all about charges which are <u>not</u> free to move. This causes them to build up in one place and it often ends with a <u>spark</u> or a <u>shock</u> when they do finally move.

Build-up of *static* is caused by *friction*

1) When two <u>insulating</u> materials are <u>rubbed</u> together, electrons will be <u>scraped off</u> one of them and <u>dumped</u> onto the other.

2) This will leave a <u>positive</u> static charge on one and a <u>negative</u> static charge on the other.

3) <u>Which way</u> the electrons are transferred depends on the <u>two materials</u> involved.

4) Electrically charged objects <u>attract</u> small objects placed near them. (Try this: rub a balloon with a cloth, then put the balloon near some bits of paper and watch them jump.)

5) The classic examples are <u>polythene</u> and <u>acetate</u> rods being rubbed with a <u>cloth duster</u>, as shown in the diagrams below.

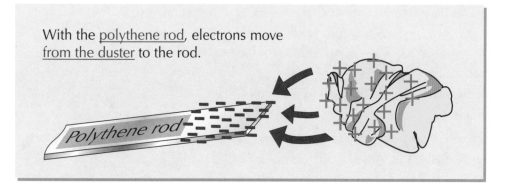

With the <u>polythene rod</u>, electrons move <u>from the duster</u> to the rod.

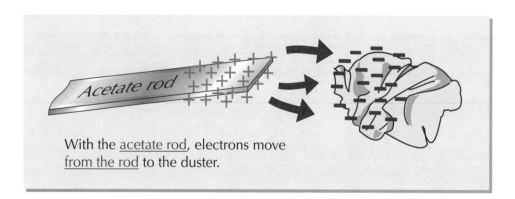

With the <u>acetate rod</u>, electrons move <u>from the rod</u> to the duster.

Static electricity is caused by electrons being transferred

Static electricity's great fun. You must have tried it — rubbing a balloon against your jumper and trying to get it to stick to the ceiling. It really works... well, sometimes. <u>Bad hair days</u> are caused by static too — it builds up on your hair, so your strands of hair repel each other. Which is nice...

Static Electricity

Only electrons move — *never the positive charges*

1) <u>Watch out for this in exams</u> — both positive and negative electrostatic charges are only ever produced by the movement of the <u>negative electrons</u>. The positive charges <u>definitely do not move</u>.

2) A <u>positive</u> static charge is always caused by <u>electrons moving away</u> elsewhere and taking their negative charges with them, as shown on the last page. Don't forget!

3) A charged conductor can be <u>discharged safely</u> by connecting it to earth with a <u>metal strap</u>.

4) The electrons flow <u>down</u> the strap to the ground if the charge is <u>negative</u> and flow <u>up</u> the strap from the ground if the charge is <u>positive</u>.

5) The <u>flow</u> of <u>electrical charge</u> is called <u>electric current</u> (see page 140).

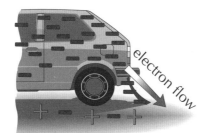 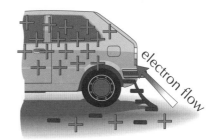

Like charges repel, *opposite* charges attract

This is <u>easy</u> and, I'd have thought, <u>kind of obvious</u>.

Two things with <u>opposite</u> electric charges are <u>attracted</u> to each other.
Two things with the <u>same</u> electric charge will <u>repel</u> each other.

These forces get <u>weaker</u> the <u>further apart</u> the two things are.

As **charge** builds up, so does the **voltage** — causing **sparks**

1) The greater the <u>charge</u> on an <u>isolated</u> object, the greater the <u>voltage</u> between it and the earth.

2) If the voltage gets <u>big enough</u> there's a <u>spark</u> which <u>jumps</u> across the gap.

3) High voltage cables can be <u>dangerous</u> for this reason. Big sparks have been known to <u>leap</u> from <u>overhead cables</u> to earth. But not often.

Static Electricity

They like asking you to give quite detailed examples in exams. Make sure you learn all these details.

Static electricity can be *dangerous*:

1) A lot of **charge** can build up on **clothes**

1) A lot of static charge can build up on clothes made out of synthetic materials if they rub against other synthetic fabrics — like when wriggling about on a car seat.

2) Eventually, this charge can become large enough to make a spark — which is really bad news if it happens near any inflammable gases or fuel fumes... KABOOM!

2) *Lightning*

Raindrops and ice bump together inside storm clouds, knocking off electrons and leaving the top of the cloud positively charged and the bottom of the cloud negative. This creates a huge voltage and a big spark.

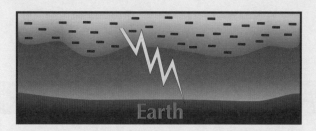

Earth

3) **Grain chutes**, **paper rollers** and the **fuel filling nightmare**

1) As fuel flows out of a filler pipe, or paper drags over rollers, or grain shoots out of pipes, then static can build up.

2) This can easily lead to a spark and might cause an explosion in dusty or fumy places — like when filling up a car with fuel at a petrol station.

3) All these problems can be solved by earthing charged objects (see next page).

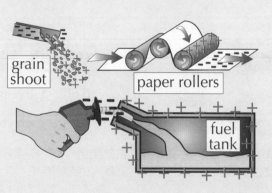

grain shoot

paper rollers

fuel tank

So it's not just bad hair days and fun with balloons then

Lightning always chooses the easiest path between the sky and the ground — even if that means going through tall buildings, trees or you. That's why it's never a good idea to fly a kite in a thunderstorm...

Static Electricity

These examples of static electricity will probably be <u>more familiar</u> to you than the ones on the previous page.

Static electricity is **annoying** *more often than it is dangerous*

1) *Attracting dust*

<u>Dust particles</u> are charged and so will be <u>attracted</u> to anything with the <u>opposite charge</u>. Unfortunately, many objects around the house are made out of <u>insulators</u> (for example, TV screen, wood, plastic containers, etc.) that get <u>easily charged</u> and attract the dust particles — this makes cleaning a <u>nightmare</u>.

2) *Clothing clings and crackles*

When <u>synthetic clothes</u> are <u>dragged</u> over each other (like in a <u>tumble drier</u>) or over your <u>head</u>, electrons get scraped off, leaving <u>static charges</u> on both parts. That leads to the inevitable — <u>attraction</u> (they stick together and cling to you) and little <u>sparks</u> or <u>shocks</u> as the charges <u>rearrange themselves</u>.

3) *Shocks from door handles*

If you walk on a <u>nylon carpet</u> wearing shoes with <u>insulating soles</u>, charge builds up on your body. Then if you touch a <u>metal door handle</u>, the charge flows to the conductor and you get a <u>little shock</u>.

Objects can be **earthed** *or* **insulated** *to* **prevent sparks**

The more dangerous problems with static electricity can be <u>avoided</u> if you're careful:

1) Dangerous <u>sparks</u> can be prevented by connecting a charged object to the <u>ground</u> using a <u>conductor</u> (e.g. a copper wire) — this is called <u>earthing</u> and it provides an <u>easy route</u> for the static charges to travel into the ground. This means <u>no charge</u> can build up to give you a shock or make a spark.

2) <u>Fuel tankers</u> must be <u>earthed</u> to prevent any sparks that might cause the fuel to <u>explode</u>.

3) Static charges are also a <u>big problem</u> in places where sparks could ignite <u>inflammable gases</u>, or where there are high concentrations of <u>oxygen</u> (e.g. in a <u>hospital</u> operating theatre).

4) <u>Antistatic sprays</u> and liquids work by making the surface of a charged object <u>conductive</u> — this provides an <u>easy path</u> for the charges to <u>move away</u> and not cause a problem.

Uses of Static Electricity

Static electricity isn't always a nuisance. It's actually got loads of applications in medicine and industry.

1) Paint sprayers — getting an even coat

1) Bikes and cars are painted using electrostatic paint sprayers.

2) The spray gun is charged, which charges up the small drops of paint. Each paint drop repels all the others, since they've all got the same charge, so you get a very fine spray.

3) The object to be painted is given an opposite charge to the gun. This attracts the fine spray of paint.

4) This method gives an even coat and hardly any paint is wasted.

5) Also, even the parts of the bicycle or car that are pointing away from the spray gun still receive paint, i.e. there are no paint shadows.

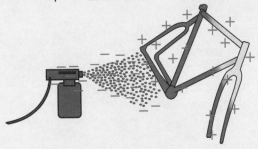

2) Dust precipitators — cleaning up emissions

Smoke is made up of tiny particles which can be removed with a precipitator. There are several different designs of precipitator — here's a very simple one:

1) As smoke particles reach the bottom of the chimney, they meet a wire grid with a high negative charge, which charges the particles negatively.

2) The charged smoke particles are attracted to positively charged metal plates. The smoke particles stick together to form larger particles.

3) When they're heavy enough, the particles fall or are knocked off the plates. They fall to the bottom of the chimney and are removed.

4) The gases coming out of the chimney contain very few smoke particles.

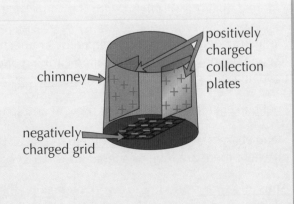

chimney

positively charged collection plates

negatively charged grid

Uses of Static Electricity

3) Defibrillators — restarting a heart

1) Hospitals and ambulances have machines called <u>defibrillators</u> which can be used to shock a stopped heart back into operation.

2) The defibrillator consists of two <u>paddles</u> connected to a power supply.

3) The paddles of the defibrillator are placed <u>firmly</u> on the patient's chest to get a <u>good electrical contact</u> and then the defibrillator is <u>charged up</u>.

4) Everyone moves away from the patient except for the defibrillator operator who holds <u>insulated handles</u>. This means <u>only the patient</u> gets a shock.

4) Photocopiers — er... copying stuff

1) The <u>image plate</u> is positively charged. An image of what you're copying is projected onto this image plate.

2) Whiter bits of the thing you're copying make <u>light</u> fall on the plate and the charge <u>leaks away</u> in those places.

3) The charged bits attract negatively charged <u>black powder</u>, which is transferred onto positively charged paper.

4) The paper is <u>heated</u> so the powder sticks.

5) Voilà, a photocopy of your piece of paper (or whatever else you've shoved in there).

6) Laserjet printers work in a similar way. Instead of an image plate, the printer has a rotating <u>image drum</u>. And the light comes from a <u>controlled laser beam</u>.

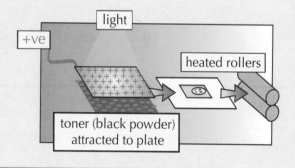

light

+ve

heated rollers

toner (black powder) attracted to plate

Clean air, beating hearts, perfect copies and brightly coloured bikes

What more do you need for a happy life? You can get your <u>very own defibrillator</u> now, to carry in your handbag just in case. (Well, it might not fit in your handbag unless you're Mary Poppins, but still...).

Warm-Up and Exam Questions

By this point you'll probably have worked out that static electricity isn't the most exciting of topics. Don't worry — there's just these few questions before you get on to much more interesting stuff.

Warm-Up Questions

1) What are the two types of electric charge?
2) Do similar charges attract or repel one another?
3) What is earthing?
4) Suggest a situation where static electricity can be dangerous.
5) What particles move when an electrically charged object is brought near another object?

Exam Questions

1   Jane hangs an uncharged balloon from a thread. She brings a negatively charged polythene rod towards the balloon.
The diagram on the right shows how the positive and negative charges in the balloon rearrange themselves when she does this.

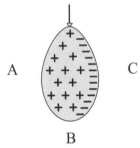

(a) In which of the positions labelled A, B and C on the diagram did Jane hold the polythene rod? Explain your answer.

(2 marks)

(b) Why are the positive charges still spread evenly over the balloon?

(1 mark)

(c) Jane brings the rod closer to the balloon. Why does the balloon swing towards it?

(2 marks)

(d) Jane keeps the rod close to the balloon, then touches the negatively charged side of the balloon with her finger.

   (i)   Describe what happens to the negative charges when she touches the balloon.

(1 mark)

   (ii) Will this leave the balloon negatively charged, positively charged or neutral?

(1 mark)

2   A positive static charge builds up on a cloth when it is used to wipe a surface.

(a) Describe the movement of charged particles that gives the cloth its charge.

(1 mark)

(b) The cloth's charge makes it more effective at dusting. Why?

(1 mark)

Exam Questions

3   The office where Jyoti works has a nylon carpet.

    (a) Jyoti often wears rubber-soled shoes to work.

        (i) What effect of static electricity might happen if she touches a metal chair leg?

(1 mark)

        (ii) Explain why this would happen.

(2 marks)

    (b) Jyoti uses a device called a Network Interface Card (NIC) to allow her laptop computer to connect to the office wi-fi network. The NIC can be harmed by static electricity.

        (i) Why does the instruction manual for the NIC recommend that she touches a metal chair leg before touching the NIC?

(1 mark)

        (ii) Would touching a wooden chair leg work in the same way? Explain your answer.

(2 marks)

4   Mike is decorating a picture frame using gold leaf. Gold leaf is made of gold that has been beaten into very thin sheets. Mike has found that his brush picks up the pieces of gold leaf better if he rubs the bristles against a cloth first. Why is this?

(2 marks)

5   Defibrillators like the one shown, are used on patients who have suffered a cardiac arrest.

    (a) What are defibrillators used for?

(1 mark)

patient — defibrillator handles

    (b) What sort of material are the handles of the defibrillator made of — insulating or conducting? Why is this?

(1 mark)

    (c) Name another device that uses static electricity.

(1 mark)

6   Lightning is one of the dangerous effects of static electricity.

    (a) What causes the build-up of static electricity before lightning occurs?

(1 mark)

    (b) Describe the distribution of charge within a cloud before a lightning strike.

(1 mark)

    (c) A lightning rod is a metal spike fixed to the top of a tall building and connected to Earth by a conducting wire. Explain how lightning rods can protect a building.

(2 marks)

Circuits — the Basics

You can use a <u>test circuit</u> to work out the <u>resistance</u> of just about anything. But first, a quick reminder...

1) <u>Current</u> (measured in A) is the <u>flow</u> of electrons round the circuit, which only happens if there's a <u>voltage</u>.

2) <u>Voltage</u> (in V) is the <u>driving force</u> that pushes the current round.

3) <u>Resistance</u> (in Ω) is anything in the circuit which <u>slows the flow down</u>.

4) <u>There's a balance</u>: the <u>voltage</u> tries to <u>push</u> the current round the circuit, and the <u>resistance</u> <u>opposes</u> it — the <u>relative sizes</u> of voltage and resistance decide <u>how big</u> the current will be:

- If you <u>increase the voltage</u>, <u>more current</u> will flow.

- If you <u>increase the resistance</u>, <u>less current</u> will flow. *(or <u>more voltage</u> is needed for the <u>same current</u>).*

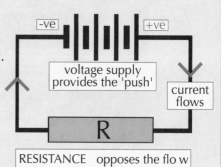

The standard *test circuit*

This is without doubt the most totally bog-standard circuit the world has ever known. So learn it.

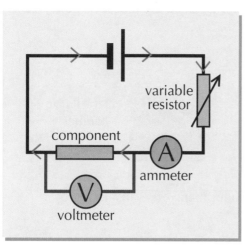

The *ammeter*

1) Measures the <u>current</u> flowing through the component.

2) Must be placed <u>in series</u> (see page 144).

3) Can be put <u>anywhere</u> in series in the <u>main circuit</u>, but <u>never</u> in parallel like the voltmeter.

The *voltmeter*

1) Measures the <u>voltage</u> across the component.

2) Must be placed <u>in parallel</u> (see page 145) around the <u>component</u> under test (<u>not</u> the variable resistor or battery!)

3) The <u>proper</u> name for 'voltage' is '<u>potential difference</u>' (<u>P.D.</u>).

Five important points

1) This <u>very basic</u> circuit is used for testing <u>components</u>, and for getting <u>V-I graphs</u> for them.

2) The <u>component</u>, the <u>ammeter</u> and the <u>variable resistor</u> are all in <u>series</u>, which means they can be put in <u>any order</u> in the main circuit. The <u>voltmeter</u>, on the other hand, can only be placed <u>in parallel</u> around the <u>component under test</u>, as shown. Anywhere else is a definite <u>no-no</u>.

3) As you <u>vary</u> the <u>variable resistor</u> it alters the <u>current</u> flowing through the circuit.

4) This allows you to take several <u>pairs of readings</u> from the <u>ammeter</u> and <u>voltmeter</u>.

5) You can then <u>plot</u> these values for <u>current</u> and <u>voltage</u> on a <u>V-I graph</u> (see next page).

The current is always shown flowing from positive to negative

The funny thing is — the <u>electrons</u> in circuits actually move from <u>–ve to +ve</u>... but scientists always think of <u>current</u> as flowing from <u>+ve to –ve</u>. Basically it's just because that's how the <u>early physicists</u> thought of it (before they found out about the electrons), and now it's become <u>convention</u>.

Resistance and Devices

I know it's dull and you've seen bits of this before, way back on page 38, but hang on in there. It's all really important stuff that can get you some nice easy marks.

Circuit symbols you should know

Here's just a quick reminder of what all the symbols mean when you're drawing a circuit:

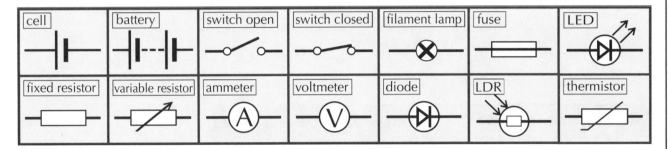

Variable resistors and diodes can alter the current in a circuit

1) A variable resistor's resistance can be changed, usually by turning a knob. The old-fashioned ones are huge coils of wire with a slider on them. They're great for altering the current flowing through a circuit. Turn the resistance up, the current drops. Turn the resistance down, the current goes up.

2) A diode is a special device made from semiconductor material such as silicon. It only lets current flow freely through it in one direction, (i.e. there's a very high resistance in the reverse direction). This turns out to be very useful in various electronic circuits.

The very important diode voltage-current graph

You've already seen the V-I graphs for filament lamps and fixed resistors (page 38) — now here's another beauty for you to learn.

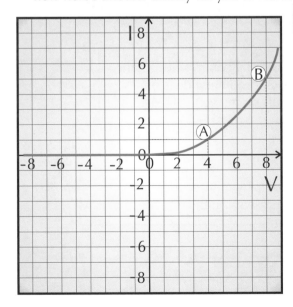

Diode

1) Current will only flow through a diode in one direction, as shown in the graph.

2) You can find R for any point by taking the pair of values (V, I) from the graph and sticking them in... you've guessed it... R = V / I (see page 39). Easy.

$$\text{resistance} = \frac{\text{potential difference}}{\text{current}}$$

3) For example — for this diode:

At point A, V = 4 V, I = 1 A, so R = 4 ÷ 1 = 4 ohms.

At point B, V = 8 V, I = 5 A, so R = 8 ÷ 5 = 1.6 ohms.

So the resistance of the diode changes.

4) You can actually tell the resistance is changing without having to get your calculator out. The graph is curved so the gradient is changing. The resistance is the inverse of the gradient or '1/gradient' (see p.38), so it must be changing too. (Remember, a straight V-I graph means a fixed resistance.)

Measuring AC

There are two different kinds of electricity supply, AC (alternating current) and DC (direct current).

Mains supply is **AC**, battery supply is **DC**

1) The UK mains is an AC supply (alternating current), which means the current is constantly changing direction (at a frequency of 50 Hz).

2) By contrast, cells and batteries supply direct current (DC). This just means that the current keeps flowing in the same direction.

AC can be shown on an **oscilloscope** screen

1) A cathode ray oscilloscope (CRO) is basically a snazzy voltmeter.

2) If you plug an AC supply into an oscilloscope, you get a 'trace' on the screen that shows how the voltage of the supply changes with time.

3) The trace goes up and down in a regular pattern — some of the time it's positive and some of the time it's negative.

4) The vertical height of the trace at any point shows the input voltage at that point.

5) There are dials on the front of the oscilloscope called the TIMEBASE and the GAIN. You can use these to set the scales of the horizontal and vertical axes of the display.

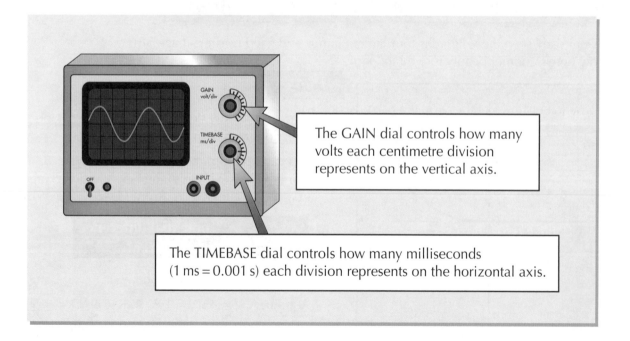

The GAIN dial controls how many volts each centimetre division represents on the vertical axis.

The TIMEBASE dial controls how many milliseconds (1 ms = 0.001 s) each division represents on the horizontal axis.

Measuring AC

Learn how to **read** an **oscilloscope trace**

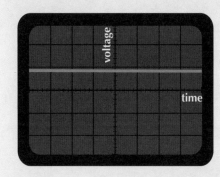

A <u>DC source</u> is always at the <u>same voltage</u>, so you get a <u>straight line</u>.

An <u>AC source</u> gives a <u>regularly repeating wave</u>. From that, you can work out the <u>period</u> and the <u>frequency</u> of the supply.

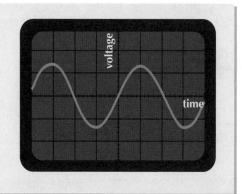

You work out the <u>frequency</u> using:

$$\text{frequency (Hz)} = \frac{1}{\text{time period (s)}}$$

EXAMPLE:

The trace shown comes from an oscilloscope with the timebase set to 5 ms/div. Find:

a) the time period   b) the frequency of the AC supply.

<u>Answer</u>:

a) To find the time period, measure the horizontal distance between two peaks.
The time period of the signal is 6 divisions. Multiply by the timebase:
Time period = 5 ms × 6 = 0.005 s × 6 = <u>0.03 s</u>

b) Using the frequency formula: Frequency = 1/0.03 = <u>33 Hz</u>

Be prepared to use traces like these to do calculations

AC's a bit harder to get your head round than DC, but it's got its advantages — especially for mains power. With AC, you can use a <u>transformer</u> to change the voltage and current, which can give more efficient transfer of electricity through the National Grid. Transformers don't work with DC currents.

Series Circuits

You need to be able to tell the difference between series and parallel circuits <u>just by looking at them</u>. You also need to know the <u>rules</u> about what happens with both types. Read on.

Series circuits — all or nothing

1) In <u>series circuits</u>, the different components are connected <u>in a line</u>, <u>end to end</u>, between the positive and negative of the power supply. (Except for <u>voltmeters</u>, which are always connected <u>in parallel</u>, but they don't count as part of the circuit).

2) If you remove or disconnect <u>one</u> component, the circuit is <u>broken</u> and they all <u>stop</u>.

3) This is generally <u>not very handy</u>, and in practice <u>very few things</u> are connected in series.

1) Potential difference is **shared**:

In series circuits the <u>total P.D.</u> of the <u>supply</u> is <u>shared</u> between the various <u>components</u>. So the <u>voltages</u> round a series circuit <u>always add up</u> to equal the <u>source voltage</u>:

$$V = V_1 + V_2 + V_3$$

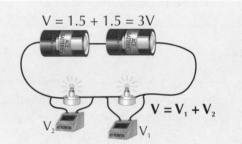

$$V = 1.5 + 1.5 = 3V$$

$$V = V_1 + V_2$$

V_2 V_1

2) Current is the **same** everywhere:

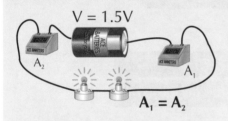

V = 1.5V

A_2

A_1

$A_1 = A_2$

1) In series circuits the <u>same current</u> flows through <u>all parts</u> of the circuit, i.e.: $\boxed{A_1 = A_2}$

2) The <u>size</u> of the current is determined by the <u>total P.D.</u> of the cells and the <u>total resistance</u> of the circuit, i.e. $I = V/R$

3) Resistance **adds up**:

1) In series circuits the <u>total resistance</u> is just the <u>sum</u> of all the resistances: $\boxed{R = R_1 + R_2 + R_3}$

2) The <u>bigger</u> the <u>resistance</u> of a component, the bigger its <u>share</u> of the <u>total P.D.</u>

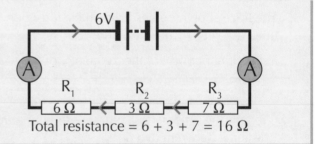

6V

A   A

R₁   R₂   R₃

6 Ω   3 Ω   7 Ω

Total resistance = 6 + 3 + 7 = 16 Ω

4) Cell voltages **add up**:

1) There is a bigger potential difference when more cells are in series, provided the cells are all <u>connected</u> the <u>same way</u>.

2) For example, when two batteries of voltage 1.5 V are <u>connected in series</u> they supply 3 V <u>between them</u>.

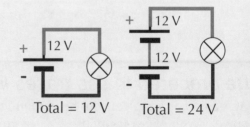

+ 12 V

+ 12 V

+ 12 V

- 12 V

-

Total = 12 V   Total = 24 V

Parallel Circuits

Parallel circuits are much more <u>sensible</u> than series circuits and so are much more <u>common</u> in <u>real life</u>.

Parallel circuits — *independence* and *isolation*

1) In <u>parallel circuits</u>, each component is <u>separately</u> connected to the +ve and –ve of the <u>supply</u>.

2) If you remove or disconnect <u>one</u> of them, it will <u>hardly affect</u> the others at all.

3) This is <u>obviously</u> how <u>most</u> things must be connected, for example in <u>cars</u> and in <u>household electrics</u>. You have to be able to switch everything on and off <u>separately</u>.

1) P.D. is the **same** across all components:

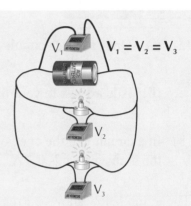

1) In parallel circuits <u>all</u> components get the <u>full source P.D.</u>, so the voltage is the <u>same</u> across all components:

$$V_1 = V_2 = V_3$$

2) This means that <u>identical bulbs</u> connected in parallel will all be at the <u>same brightness</u>.

2) Current is **shared** between branches:

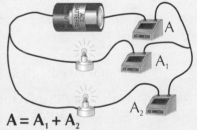

1) In parallel circuits the <u>total current</u> flowing around the circuit is equal to the <u>total</u> of all the currents in the <u>separate branches</u>. $A = A_1 + A_2 + A_3$

2) In a parallel circuit, there are <u>junctions</u> where the current either <u>splits</u> or <u>rejoins</u>. The total current going <u>into</u> a junction has to equal the total current <u>leaving</u> it.

$$A = A_1 + A_2$$

3) If two <u>identical components</u> are connected in parallel, the <u>same current</u> flows through each one.

3) Resistance is **tricky**:

1) The <u>current</u> through each component depends on its <u>resistance</u>. The <u>lower</u> the resistance, the <u>bigger</u> the current that'll flow through it.

2) The <u>total resistance</u> of the circuit is <u>tricky to work out</u>, but it's always <u>less</u> than that of the branch with the <u>smallest</u> resistance.

Voltmeters and ammeters are **exceptions** to the rule:

1) Ammeters and voltmeters are <u>exceptions</u> to the series and parallel rules.

2) Ammeters are <u>always</u> connected in <u>series</u> even in a parallel circuit.

3) Voltmeters are <u>always</u> connected in <u>parallel with a component</u> even in a series circuit.

Warm-Up and Exam Questions

Phew — circuits aren't the easiest thing in the world, are they? Make sure you've understood them though, by trying these questions. If you get stuck, just go back and reread the relevant page.

Warm-Up Questions

1) What are the units of resistance?
2) If the resistance of a circuit is increased, what happens to the current?
3) Draw the symbol for a light-emitting diode (LED).
4) What is the frequency of UK mains electricity supply?
5) What is the time period of a wave with a frequency of 100 Hz?

Exam Questions

1 (a) (i) Draw a series circuit containing a cell, a filament lamp and an ammeter.

(3 marks)

(ii) Add arrows to your diagram showing the direction of conventional current.

(1 mark)

(b) In a circuit like the one you have drawn, the resistance of the filament lamp is 5 Ω and the ammeter reads 0.3 A. Calculate the potential difference across the cell.

(1 mark)

2 The graph below shows current against potential difference (p.d.) for a diode.

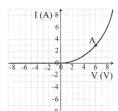

(a) Explain why the graph shows zero current for negative p.d.s?
(1 mark)

(b) What is the resistance of the diode at the point marked A?
(2 marks)

3 The diagram on the right shows a trace on a CRO.

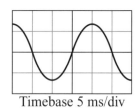

Timebase 5 ms/div

(a) Is the trace displaying the output from the mains or a battery? Give a reason for your answer.
(1 mark)

(b) What is the time period of the wave?
(1 mark)

(c) What is the frequency of the wave?
(1 mark)

(d) What will happen to the CRO trace if the voltage of the supply is reduced?
(1 mark)

4 A parallel circuit is connected as shown. Complete the readings on the voltmeters and ammeters on the diagram.

(5 marks)

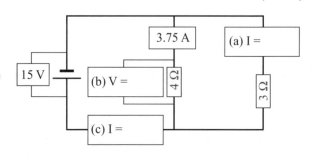

Fuses and Safe Plugs

Electricity is dangerous. Just watch out for it, that's all.

Electrical cables usually have *live*, *neutral* and *earth* wires

The <u>brown</u> wire is the <u>LIVE WIRE</u>.
This wire alternates between a
<u>HIGH +VE AND −VE VOLTAGE</u>
of about <u>230 V</u>.

The <u>blue</u> wire is the <u>NEUTRAL WIRE</u>.
This wire is always at <u>0 V</u>.

Electricity normally flows in through the
<u>live</u> wire and out through the <u>neutral</u> wire.

The <u>green and yellow</u> wire is the <u>EARTH WIRE</u>.

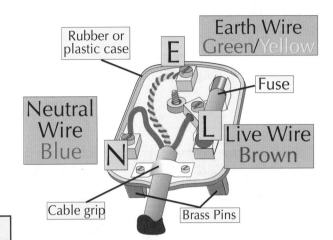

Rubber or plastic case

Earth Wire Green/Yellow

E

Fuse

Neutral Wire Blue

N L

Live Wire Brown

Cable grip

Brass Pins

Learn the *safety features* of *plugs* and *cables*

Get the *wiring right:*

1) The <u>right coloured wire</u> is connected to each pin, and <u>firmly screwed</u> in.
2) <u>No bare wires</u> are showing inside the plug.
3) The <u>cable grip</u> is tightly fastened over the cable <u>outer layer</u>.

Plug features:

1) The <u>metal parts</u> are made of copper or brass because these are <u>very good conductors</u>.
2) The case, cable grip and cable insulation are made of <u>rubber</u> or <u>plastic</u> because they're really good <u>insulators</u>, and <u>flexible</u> too.

Residual Current Circuit Breakers help *prevent electrocution*

Sometimes, you can protect yourself with a <u>Residual Current Circuit Breaker</u> (<u>RCCB</u>) instead of a fuse and an earth wire. RCCBs work slightly differently.

1) Normally exactly the <u>same current</u> flows through the <u>live</u> and <u>neutral</u> wires.
2) However, if somebody <u>touches</u> the live wire, a very large current will flow <u>through them</u> to the <u>earth</u>. So now the neutral wire is carrying <u>less current</u> than the live wire.
3) The RCCB detects this <u>difference</u> in current and <u>quickly cuts off the power</u>.
4) An RCCB can easily be <u>reset</u> by flicking a switch on the device. This makes them <u>more convenient</u> than fuses — which have to be <u>replaced</u> once they've melted.

Fuses and Safe Plugs

Earthing and fuses prevent fires and electric shocks

The earth wire doesn't normally have any current flowing through it.
The earth wire and fuse are just there for safety — and they work together like this:

1) If a fault develops in which the live somehow touches the metal case, then because the case is earthed, a big current flows in through the live, through the case and out down the earth wire.

2) This surge in current 'blows' the fuse (or trips the circuit breaker), which cuts off the live supply.

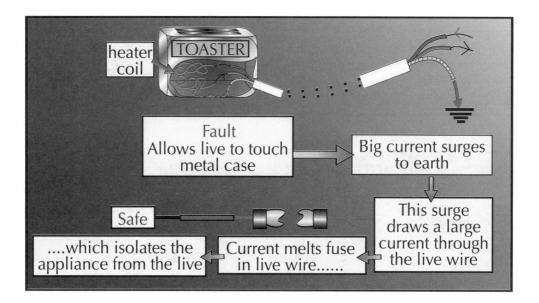

3) This isolates the whole appliance, making it impossible to get an electric shock from the case.

4) It also prevents the risk of fire caused by the heating effect of a large current.

5) Fuses should be rated as near as possible but just higher than the normal operating current.

All appliances with metal cases must be "earthed" to reduce the danger of electric shock.
"Earthing" just means the case must be attached to an earth wire.
An earthed conductor can never become live.

If the appliance has a plastic casing and no metal parts exposed, it's said to be double insulated. Anything with double insulation just needs a live and a neutral wire.

Fuses — you'll find them in exams and kettles

It's really important to learn the colour of each of the wires — not just for the exam but for real life too.
I like to remember them like this... green is for grass, and where does grass grow — in the earth of course. Now blue and neutral both have e and u in so they must go together, which just leaves the live wire — if it's alive it must be a worm, so is brown. Easy peasy... ish.

Energy and Power in Circuits

You can look at <u>electrical circuits</u> in <u>two ways</u>. The first is in terms of a voltage <u>pushing the current</u> round and the resistances opposing the flow, as on page 140. The <u>other way</u> of looking at circuits is in terms of <u>energy transfer</u>. Learn them <u>both</u> and be ready to tackle questions about <u>either</u>.

Energy is *transferred* from cells and other *sources*

Anything which <u>supplies electricity</u> is also supplying <u>energy</u>.

So cells, batteries, generators, etc. all <u>transfer energy</u> to components in the circuit:

| <u>Motion</u>: motors | <u>Light</u>: light bulbs | <u>Heat</u>: Hairdriers/kettles | <u>Sound</u>: speakers |

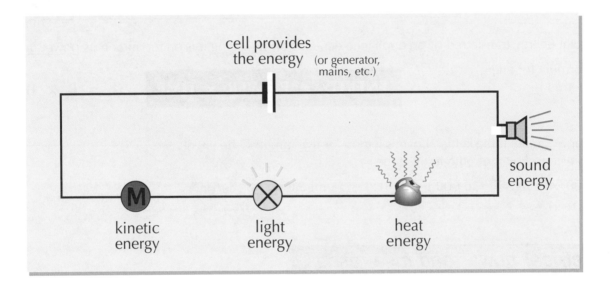

cell provides
the energy (or generator, mains, etc.)

kinetic energy   light energy   heat energy   sound energy

All resistors produce heat when a current flows through them

1) Whenever a <u>current</u> flows through anything with <u>electrical resistance</u> (which is pretty well everything) then <u>electrical energy</u> is converted into <u>heat energy</u>.

2) The <u>more current</u> that flows, the more heat is produced.

3) A <u>bigger voltage</u> means more heating because it pushes more current through.

4) You can <u>measure</u> the amount of heat produced by putting a resistor in a known amount of water, or inside a solid block, and measuring the increase in temperature.

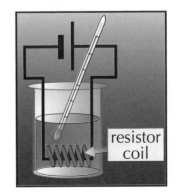

resistor coil

Energy and Power in Circuits

Power ratings tell you how much energy a device transfers per second.

Power ratings of appliances

A light bulb converts <u>electrical energy</u> into <u>light</u> and has a power rating of 100 watts (W), which means it transfers <u>100 joules/second</u>.

A kettle converts <u>electrical energy</u> into <u>heat</u> and has a power rating of 2.5 kW, transferring <u>2500 joules/second</u>.

The total energy transferred by an appliance depends on <u>how long</u> it is on for and on its <u>power rating</u>.

The formula for energy transferred is:

$$\textbf{ENERGY = POWER} \times \textbf{TIME} \qquad \textbf{(E = P} \times \textbf{t)}$$

<u>For example</u>, if the kettle above is left on for <u>five minutes</u>, the energy transferred by the kettle in this time is:

300 × 2500 = 750 000 J = <u>750 kJ</u>.   (5 minutes = 300 seconds)

Electrical power and fuse ratings

1) The formula for <u>electrical power</u> is:

$$\textbf{POWER = VOLTAGE} \times \textbf{CURRENT} \qquad \textbf{(P = V} \times \textbf{I)}$$

2) Most electrical goods show their <u>power rating</u> and <u>voltage rating</u>. To work out the <u>fuse</u> needed, you need to work out the <u>current</u> that the item will normally use:

Example:

A hairdrier is rated at 230 V, 1 kW. Find the fuse needed.

<u>Answer:</u>   I = P/V = 1000/230 = 4.3 A.

Normally, the fuse should be rated just a bit higher than the normal current, so a 5 amp fuse is ideal for this one.

Don't just pick the nearest fuse value — a lower one is no good

In the UK, you can usually get fuses rated at <u>3 A</u>, <u>5 A</u> or <u>13 A</u>, and that's about it. You should bear that in mind when you're working out fuse ratings. If you find you need a 10.73 A fuse — tough.

Charge, Voltage and Energy

Total charge through a circuit depends on current and time

1) Current is the <u>flow of electrical charge</u> (in coulombs, C) around a circuit.

2) When a <u>current</u> (I) flows past a point in a circuit for a <u>time</u> (t) then the <u>charge</u> (Q) that has passed is given by:

total charge = current × time

3) <u>More charge</u> passes around the circuit when a <u>bigger current</u> flows.

Example:

A battery charger passes a current of 2.5 A through a cell for a period of 4 hours. How much charge does the charger transfer to the cell in total?

<u>Answer:</u> Q = I × t = 2.5 × (4 × 60 × 60) = 36 000 C (36 kC).

Voltage is the energy transferred per charge passed

1) When an electrical <u>charge</u> (Q) goes through a <u>change</u> in voltage (V), then <u>energy</u> (E) is <u>transferred</u>.

2) Energy is <u>supplied</u> to the charge at the <u>power source</u> to 'raise' it through a voltage.

3) The charge <u>gives up</u> this energy when it '<u>falls</u>' through any <u>voltage drop</u> in <u>components</u> elsewhere in the circuit.

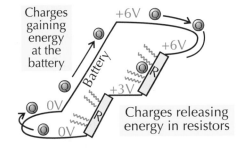

The formula is really simple:

energy transformed = charge × potential difference

4) The <u>bigger</u> the <u>change</u> in voltage (or P.D.), the <u>more energy</u> is transferred for a <u>given amount of charge</u> passing through the circuit.

5) That means that a battery with a <u>bigger voltage</u> will supply <u>more energy</u> to the circuit for every <u>coulomb</u> of charge which flows round it, because the charge is raised up 'higher' at the start (see above diagram) — and as the diagram shows, <u>more energy</u> will be <u>dissipated</u> in the circuit too.

Example:

A motor is attached to a 3 V battery. If a current of 0.8 A flows through the motor for 3 minutes:

a) Calculate the total charge passed.

b) Calculate the energy transformed by the motor.

c) Explain why the kinetic energy output of the motor will be less than your answer to b).

<u>Answer:</u> a) Using the formula above, Q = I × t = 0.8 × (3 × 60) = <u>144 C</u>.

b) Use E = Q × V = 144 × 3 = <u>432 J</u>.

c) The motor won't be 100% efficient. Some energy will be transformed into <u>sound and heat</u>.

Warm-Up and Exam Questions

Yep, it's the end of Section Five. Check you can do the straightforward stuff with this warm-up, then have a go at the exam questions below.

Warm-Up Questions

1) Which of the live, neutral or earth wires is always at 0 volts?
2) What energy transformation occurs when electric current flows through a resistor?
3) What is the name for the rate at which electrical charge flows round a circuit?
4) What is the equation linking Q, V and E?
5) Which of the earth, live, or neutral wires is not used in a double-insulated device?

Exam Questions

1   (a)   What colour are each of the following wires in an electric plug?

      (i)   live

      (ii) neutral

      (iii) earth

(3 marks)

   (b) Which two wires usually carry the same current?

(1 mark)

   (c) What type of safety device contains a wire that is designed to melt when the current passing through it goes above a certain value?

(1 mark)

2   A domestic appliance has a plug containing live, neutral and earth wires and a fuse, all correctly wired to the household circuit.

   (a) Describe the current that flows in each of the wires in the following situations:

      (i)   normal operation

      (ii) the instant after the live wire has come into contact with the metal cover

      (iii) after the fuse has blown

(3 marks)

   (b) If there were no earth wire or fuse present and a person touched the live wire, what path would the current take?

(1 mark)

   (c) Explain how a residual current circuit breaker works.

(2 marks)

3   A current of 0.5 A passes through a torch bulb. The torch is powered by a 3 V battery.

   (a) What is the power of the torch?

(2 marks)

   (b) If the torch is on for half an hour, how much charge has passed through the battery?

(2 marks)

   (c) How much electrical energy does the bulb transfer in half an hour?

(2 marks)

Revision Summary for Section Five

There's some pretty heavy physics in this section. But just take it one page at a time and it's not so bad. When you think you know it all, try these questions and see how you're getting on. If there are any you can't do, look back at the right bit of the section, then come back here and try again.

1)   What causes the build-up of static electricity? Which particles move when static builds up?

2)   Give three examples each of static electricity being: a) a nuisance, b) dangerous.

3)   Explain what's done to avoid the build-up of static electricity, and how the method works.

4)   Give four examples of how static electricity can be helpful. Describe each example in detail.

5)   Explain what current, voltage and resistance are in an electric circuit.

6)   Sketch typical voltage-current graphs for a diode. Explain the shape of the graph.

7)* An oscilloscope is plugged into the mains (50 Hz). Sketch what you would expect to see on the screen if the timebase is set to 2 ms/div.

8)* Find each unknown voltage, current or resistance in this circuit.

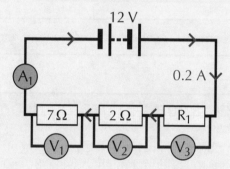

9)   Why are parallel circuits often more useful than series ones?

10)*Find each unknown voltage, current or resistance in this circuit.

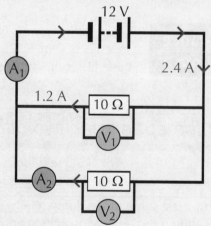

11) Sketch a properly wired three-pin plug.

12) Explain fully how a fuse and earth wire work together.

13) What is double insulation?

14)*Find the appropriate fuse (3 A, 5 A or 13 A) for each of these appliances:

    a)   a toaster rated at 230 V, 1100 W,

    b)   an electric heater rated at 230 V, 2000 W.

15)*Find the total charge passed when a current of 3 A flows for 1 hour 20 minutes.

16)*Calculate the energy transformed by a torch using a 6 V battery when 530 C of charge pass through.

                                                          * Answers on page 275.

Relative Speed and Velocity

When you're talking about the motion of a car, its <u>direction</u> is important as well as its <u>speed</u>.

Speed is just a number, but velocity has direction too

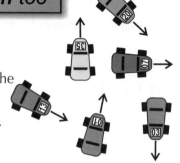

1) To measure the <u>speed</u> of an object, you only need to measure <u>how fast</u> it's going — the <u>direction</u> is <u>not important</u>. E.g. speed = 30 mph.

2) <u>Velocity</u> is a <u>more useful</u> measure of <u>motion</u>, because it describes both the <u>speed and direction</u> (see page 102). E.g. velocity = 30 mph due north.

3) A quantity like <u>speed</u>, that has only a <u>number</u>, is called a <u>scalar</u> quantity. A quantity like <u>velocity</u>, that has a <u>direction as well</u>, is a <u>vector</u> quantity.

> <u>Scalar quantities</u>: mass, time, temperature, length, etc.

> <u>Vector quantities</u>: force, displacement, acceleration, momentum, etc.

All speeds = 0.5 m/s
Velocities = completely different

You can calculate average speed to get the overall picture

1) <u>Speed</u> is a measure of <u>how fast</u> an object is moving at a given <u>moment</u>, but over a journey your speed <u>changes</u>. E.g. if I drove from Chester to London, I might drive at 30 miles per hour leaving Chester, then get up to 70 mph on the motorway, then slow down to 2 mph to find a parking place in London.

2) You know that speed = distance ÷ time (see page 102). So you can calculate the <u>average speed</u> of a journey, where the <u>distance</u> is the <u>total distance</u> covered, and the <u>time</u> is the <u>total time</u> taken:

> **average speed** $=$ $\dfrac{\textbf{total distance}}{\textbf{total time}}$

E.g. a walker covers 600 m in 375 s. His average speed is 600 m ÷ 375 s = 1.6 m/s... but he's probably gone faster or slower than that during the walk.

Relative speed compares the speeds of two different objects

1) When you look out of a <u>car window</u>, a car that's <u>overtaking</u> you looks like it's <u>not moving very fast</u>. Whereas a car on the <u>opposite side</u> of the motorway seems to <u>whizz past</u> at 100 miles an hour.

2) It's all to do with <u>relative speed</u> — how fast something's going <u>relative to something else</u>. The easiest way to think of it is to imagine yourself in a moving car, watching another vehicle from the window.

3) A car going the <u>same way</u> as you will only have a small speed <u>relative to your car</u>, whereas a car going the <u>opposite way</u> will have a much <u>bigger</u> speed <u>relative to you</u>.

<u>Car travelling the same way as you</u>

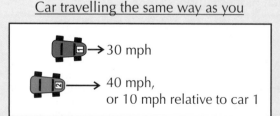

<u>Car travelling the opposite way to you</u>

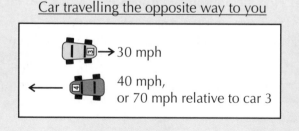

Combining Velocities and Forces

If a bird has the <u>wind behind it</u>, it flies a bit <u>faster</u>. Likewise, if it's flying <u>into the wind</u>, it'll be slower.
To work out the velocity as seen by a '<u>stationary observer</u>' (someone <u>standing still</u>), you have to <u>combine</u>
the <u>velocity of the bird</u> with the <u>velocity of the wind</u>. Get ready for a bit of <u>maths</u> — this is vector stuff...

To **combine two vectors**, you add them **end to end**

1) **With or against** the **current** — **EASY**

It's <u>easy</u> when the plane (or whatever) is flying <u>directly into the wind</u>
(or whatever), or with the <u>wind behind it</u>. For example:

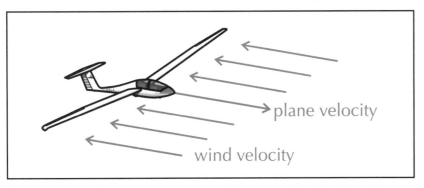

On the <u>vector diagrams</u> you just need arrows going back and forwards, like this:

Example:

A light plane is flying east. Its airspeed indicator shows 120 km/h. It is flying into a
wind of 20 km/h — i.e. within a stream of air that's moving west at 20 km/h.

What is its resultant velocity?

Draw the vectors <u>end to end</u>:

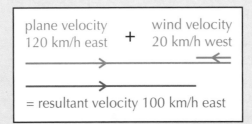

So an observer on the ground would see the plane going <u>east at 100 km/h</u>.

Of course, it's not always that straightforward

As you'll see on the next page, the velocities <u>won't</u> always be conveniently acting in opposite
directions, and you need to know how to deal with that. But once you've got the hang of it, you can
use the same trick to combine <u>any vectors</u> — momentum, displacement, acceleration, anything.
Just draw the vectors end to end and, with a bit of maths, you can find the overall (resultant) vector.

Combining Velocities and Forces

Here's where combining velocities gets a little bit more <u>complicated</u>...

2) Across the current — a bit more maths

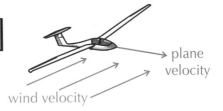

If the <u>plane</u> (or whatever) is flying <u>across the wind</u> (or whatever), it's a bit <u>more tricky</u>.

Example:

> A boat is going west at 14 m/s (according to the speed indicator) in a river with a current running north at 8 m/s. What is its resultant velocity?

Again, you draw the vectors <u>end to end</u>, only this time it makes a triangle:

To work out the resultant velocity, you need both speed and direction. It's a right-angled triangle, so:

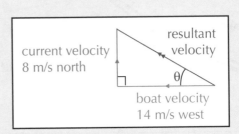

For <u>speed</u> you need <u>Pythagoras'</u> theorem:

And for <u>direction</u>, it's good old <u>trigonometry</u>:

$$\text{speed} = \sqrt{(8^2 + 14^2)} = \underline{16.1 \text{ m/s}}$$

$$\tan \theta = 8/14, \text{ so } \theta = \tan^{-1} (8/14) = \underline{29.7°}$$

It's the same with forces and ANY vectors at right angles

Example:

> Two horses pull a boat along a canal, with forces at <u>right angles</u> to each other.

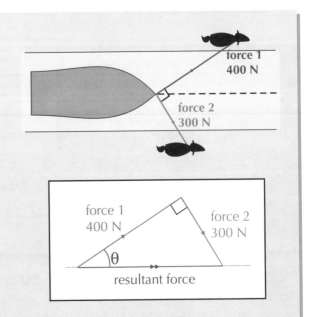

Draw the vectors <u>end to end</u>, to make a right-angled triangle:

And it's Pythagoras again:

Size of force = $\sqrt{(300^2 + 400^2)}$ = <u>500 N</u>

Direction = angle of θ to force 1, which you find by trig:

$\tan \theta = 300/400$, so $\theta = \tan^{-1} (3/4) = \underline{36.9°}$

Equations of Motion

These equations of motion are dead handy for working out velocity, acceleration and other goodies...

You need to know these *four equations of motion*

Which of these equations you need to use depends on what you already know, and what you need to find out. But that means you have to know all four equations — preferably like the back of your hand.

$$s = \frac{(u + v)}{2} t$$

$$v = u + at$$

$$s = ut + \tfrac{1}{2} at^2$$

$$v^2 = u^2 + 2as$$

Altogether, there are five things involved in these equations:

$\underline{u}$ = initial velocity, $\underline{v}$ = final velocity,

$\underline{s}$ = displacement, $\underline{a}$ = acceleration, $\underline{t}$ = time.

Make sure you use the *right equation*

If you know three things, you can find out either of the other two — if you use the right equation, that is. And if you use this method twice, you can find out both things you don't know.

HOW TO CHOOSE YOUR EQUATION:

1)   Write down which three things you already know.

2)   Write down which of the other things you want to find out.

3)   Choose the equation that involves all the things you've written down.

4)   Stick in your numbers, and do the maths.

REMEMBER:
Direction's important for velocity, acceleration and displacement —
always decide which direction is positive, and then stick with it.

Equations of Motion

Don't worry — I wouldn't let you go off into the big wide world armed only with four equations. Here are some <u>examples</u> to show you just how they should be used.

Example 1

A car going at 10 m/s accelerates at 2 m/s² for 8 seconds. How far does the car travel while it is accelerating?

Now, first things first... I'll say that the '<u>positive</u>' direction is '<u>to the right</u>'.

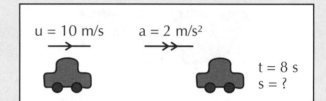

1) You know <u>u</u> (= 10 m/s), <u>a</u> (= 2 m/s²) and <u>t</u> (= 8 s).

2) You want to <u>find out</u> the distance, <u>s</u>.

3) So you need the equation with <u>all</u> these in: <u>u</u>, <u>a</u>, <u>t</u> and <u>s</u> — the 3rd equation: <u>s = ut + ½at²</u>.

4) Put the numbers in: s = (10 × 8) + ½ (2 × 8²) = 80 + 64 = <u>144 m</u>.

Example 2

A car going at 25 m/s decelerates at 1.5 m/s² as it heads towards a built-up area that is 145 m away. What will its velocity be when it reaches the built-up area?

I'll make the '<u>positive</u>' direction '<u>to the right</u>' again.

1) You know <u>u</u> (= 25 m/s), <u>a</u> (= −1.5 m/s² ...don't forget it's negative!) and <u>s</u> (= 145 m).

u = 25 m/s   a = −1.5 m/s²

s = 145 m
v = ?

2) You want to <u>find out</u> the final velocity, <u>v</u>.

3) So you need the equation with <u>all</u> these in: <u>u</u>, <u>a</u>, <u>s</u> and <u>v</u> — that will be the fourth equation: <u>v² = u² + 2as</u>.

4) Put the numbers in: v² = 25² + 2(−1.5) (145) = 190, so v = $\sqrt{190}$ = <u>13.8 m/s</u>

Learn the four equations and what the five letters mean

It's no good knowing the <u>equations</u> forwards, backwards and upside down if you can't <u>relate</u> them to the <u>question</u> because you've forgotten what the letters mean. Motion questions can be tricky at first, but that method works for all of them. <u>Practise</u> the examples — cover up my working and do them yourself.

Projectile Motion

Exciting stuff, this — things flying through the air, where the only force on them is due to <u>gravity</u>.

The **path** of a **projectile** is **always a parabola**

1) A <u>projectile</u> is something that is <u>projected</u>, or <u>thrown</u>, and from then on has <u>only gravity</u> acting on it (ignoring air resistance).

2) So things like cannonballs, golf balls, paper darts and long-jumpers are all projectiles.

3) The <u>path</u> a projectile takes through the air (called its <u>trajectory</u>) is always a <u>parabola</u>, which is this shape:

Deal with **horizontal** and **vertical** motion **separately**

1) Motion can be split into <u>two</u> separate bits — the <u>horizontal</u> bit and the <u>vertical</u> bit.

2) These bits are totally <u>separate</u> — one doesn't affect the other.

3) So gravity (which only acts downwards) <u>doesn't affect horizontal motion at all</u>.

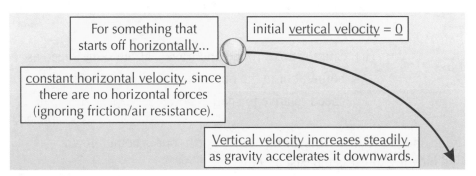

For something that starts off <u>horizontally</u>...

initial <u>vertical velocity = 0</u>

<u>constant horizontal velocity</u>, since there are no horizontal forces (ignoring friction/air resistance).

<u>Vertical velocity increases steadily</u>, as gravity accelerates it downwards.

4) Both bits of the motion — the horizontal velocity and the vertical velocity — are <u>vectors</u>. The overall (resultant) velocity of the ball at any point is the <u>vector sum</u> of the separate bits (see p.156).

Projectile **calculations** use the **equations of motion**

Example

A football is kicked horizontally from a 20 m wall.

a) How long is it before it lands? (use $g = 10$ m/s².)

b) If its horizontal velocity is originally 5 m/s, how far does it travel before it lands?

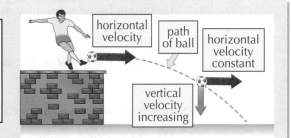

horizontal velocity

path of ball

horizontal velocity constant

vertical velocity increasing

a) It lands when it's travelled 20 m vertically. Using $s = ut + \frac{1}{2}at^2$, where $u = 0$, $a = 10$ m/s², $s = 20$:

$20 = (0 \times t) + \frac{1}{2}at^2 = 10t^2 / 2$, i.e. $\underline{t = 2 \text{ s when it lands}}$.

b) Using distance = speed × time, where $v = 5$ and $t = 2$ (from above): $\underline{s = 5 \times 2 = 10 \text{ m}}$.

Warm-Up and Exam Questions

This section has far too many formulas for its own good — make sure you know when and how to use each one. There's plenty of practice here for you.

Warm-Up Questions

1) What's the difference between vector and scalar quantities?
2) Name two vector and two scalar quantities.
3) How would you work out the relative speed of two objects moving in the same direction?
4) Write down two different equations of motion for displacement.
5) What forces act on a projectile?

Exam Questions

1 (a) Janine walks 192 m to her car, taking 2 minutes, and then drives 3600 m in 3 minutes. What is her average speed in m/s?

(2 marks)

(b) The diagram below shows the motion of two snails, A and B.

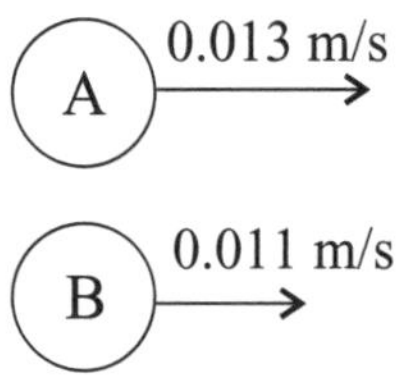

A   0.013 m/s →

B   0.011 m/s →

(i) What is the speed of snail A relative to snail B?

(1 mark)

(ii) Snail A turns round and travels at the same speed as before, but in the opposite direction. What is its speed relative to snail B now?

(1 mark)

2 A bird's normal flying speed is 6 m/s. It is aiming to fly north, but is being blown off course by a wind blowing towards the east at 9 m/s.

(a) Draw a vector diagram to show the resultant velocity of the bird.

(3 marks)

(b) Calculate the bird's resultant velocity.

(4 marks)

3 (a) Sarah is riding her motorbike at a velocity of 9 m/s. She rides down a hill and accelerates at a rate of 7 m/s^2 until she reaches the bottom of the hill 3 seconds later.

(i) Calculate Sarah's velocity at the bottom of the hill.

(2 marks)

(ii) Calculate the distance she covered while accelerating.

(1 mark)

(b) A ball is thrown horizontally out of an upper window. It hits the ground after 1.73 s.

(i) How high was the window? Assume no air resistance. ($g = 10$ m/s^2)

(2 marks)

(ii) The ball travels 2.59 m horizontally before it hits the ground. What was the horizontal velocity of the ball when it left the window?

(1 mark)

Turning Forces and Centre of Mass

<u>Turning forces</u> — full of spanners and levers. You should have a good idea about calculating moments, pivots and centre of mass by the time you've finished these pages.

A *moment* is the *turning effect* of a force

MOMENT (Nm) = FORCE (N) × perpendicular DISTANCE (m)
(between line of action and pivot)

1) The <u>force</u> on the spanner causes a <u>turning effect</u> or <u>moment</u> on the nut. A <u>larger</u> force would mean a <u>larger</u> moment.

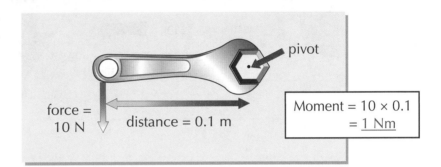

force = 10 N   distance = 0.1 m

Moment = 10 × 0.1
= <u>1 Nm</u>

2) Using a longer spanner, the same force can exert a <u>larger</u> moment because the <u>distance</u> from the pivot is <u>greater</u>.

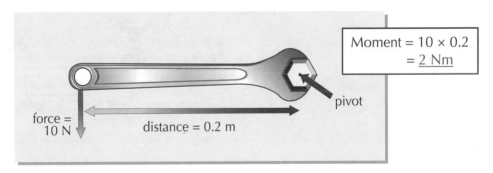

Moment = 10 × 0.2
= <u>2 Nm</u>

force = 10 N   distance = 0.2 m   pivot

3) To get the <u>maximum</u> moment (or turning effect) you need to push at <u>right angles</u> (<u>perpendicular</u>) to the spanner.

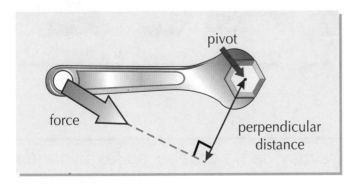

pivot

force

perpendicular distance

4) Pushing at <u>any other angle</u> means a smaller moment because the <u>perpendicular</u> distance between the line of action and the pivot is <u>smaller</u>.

Turning Forces and Centre of Mass

The centre of mass hangs *directly below* the *point of suspension*

1) You can think of the <u>centre of mass</u> of an object as the point where its <u>whole</u> mass is concentrated.

2) A freely suspended object will <u>swing</u> until its centre of mass is <u>vertically below</u> the <u>point of suspension</u>.

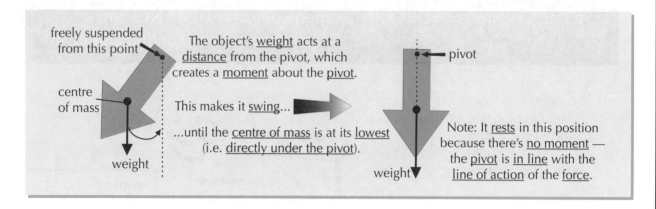

3) This means you can find the <u>centre of mass</u> of any flat shape like this:

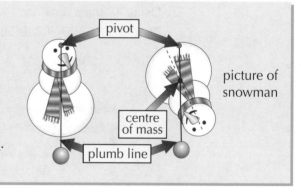

a) Suspend the shape and a <u>plumb line</u> from the same point, and wait until they <u>stop moving</u>.

b) <u>Draw</u> a line along the plumb line.

c) Do the same thing again, but suspend the shape from a <u>different</u> pivot point.

d) The centre of mass is where your two lines <u>cross</u>.

4) But you don't need to go to all that trouble for <u>simple</u> shapes. You can quickly guess where the centre of mass is by looking for <u>lines of symmetry</u>.

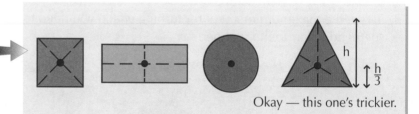

Okay — this one's trickier.

Turning forces and centre of mass may be new ideas, so take your time

Think of the extra force you need to open a door by pushing it <u>near the hinge</u> compared to <u>at the handle</u> — the <u>distance from the pivot</u> is <u>less</u>, so you need <u>more force</u> to get the <u>same moment</u>. For the centre of mass, try and get <u>loads of practice</u> finding it for different shapes, until you're sure.

Balanced Moments and Stability

Once you can calculate moments, you can work out if a <u>seesaw is balanced</u>. Useful thing, Physics.

A question of **balance** — are the **moments equal**?

If the <u>anticlockwise moments</u> are equal to the <u>clockwise moments</u>, the object <u>won't turn</u>.

Example 1

Your younger brother weighs 300 N. He sits 2 m from the pivot of a seesaw.
If you weigh 700 N, where should you sit to balance the seesaw?

For the seesaw to <u>balance</u>: **total anticlockwise moments = total clockwise moments**

anticlockwise moment = clockwise moment
$$300 \times 2 = 700 \times y$$
$$y = \underline{0.86 \text{ m}}$$

*Ignore the weight of the seesaw — its centre of mass
is on the pivot, so it doesn't have a turning effect.*

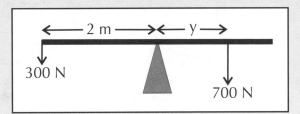

Example 2

A 6 m long steel girder weighing 1000 N rests horizontally on a pole 1 m from one end.
What is the tension in a supporting cable attached vertically to the other end?

The '<u>tension in the cable</u>' bit makes it
sound harder than it actually is.
But the girder's <u>weight</u> is <u>balanced</u> by
the tension <u>force</u> in the cable, so...

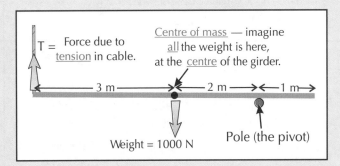

anticlockwise moment   =   clockwise moment
(due to weight)   (due to tension in cable)

$$1000 \times 2 = T \times 5$$
$$2000 = 5T$$
and so $\underline{T = 400 \text{ N}}$

Balanced Moments and Stability

And if the moments aren't equal...

If the total anticlockwise moments do not equal the total clockwise moments, there will be a resultant moment

...so the object will <u>turn</u>.

Low and wide objects are most stable

<u>Unstable</u> objects tip over easily, and <u>stable</u> objects don't. The position of the <u>centre of mass</u> is all-important in determining whether an object is stable or not.

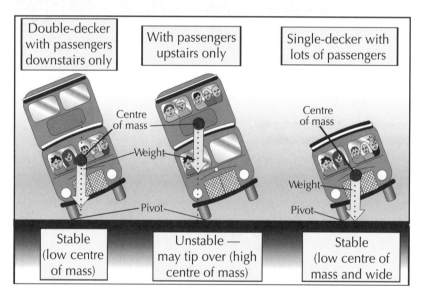

1) The most <u>stable</u> objects have a <u>wide base</u> and a <u>low centre of mass</u>.

2) An object will begin to <u>tip over</u> if its centre of mass moves <u>beyond</u> the edge of its base.

3) Again it's because of <u>moments</u> — if the weight <u>doesn't</u> act <u>in line</u> with the <u>pivot</u>, it'll cause a <u>resultant moment</u>.

4) This will either right the object or tip it over.

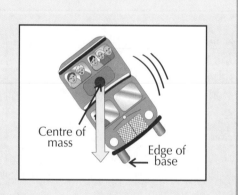

You can often tell just by looking what will tip easily and what won't

If you've got two <u>equal and opposite moments</u>, you've got <u>equilibrium</u>, and your thing-on-a-pivot will stay still. Remember that and you won't go far wrong (as long as you calculate the moments properly). <u>Learn</u> the factors that make an object hard to tip over — a <u>low centre of mass</u> and a <u>wide base</u>.

Warm-Up and Exam Questions

You know the drill — some warm-up questions to make sure you're awake, then some proper exam style questions to make sure you really know your stuff. Don't forget to check up on any tricky bits.

Warm-Up Questions

1) How do you calculate the moment of a force around a pivot?
2) What unit is used to describe the moment of a force?
3) What happens to an object if the clockwise and anticlockwise moments on it are not equal?
4) What is meant by the centre of mass of an object?
5) Why will an object begin to tip over if its centre of mass moves beyond the edge of its base?

Exam Questions

1   Bolts are often secured using an Allen key, as shown in the diagram. One end of the allen key is put into the bolt and the other is turned to tighten the bolt.

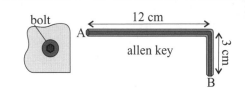

(a) Calculate the moment on the bolt when:

    (i)   End A of the Allen key is put into the bolt and a force of 15 N is applied to end B.

(2 marks)

    (ii) End B of the Allen key is put into the bolt and a force of 15 N is applied to end A.

(2 marks)

(b) Which end of the allen key (A or B) should be put into the bolt to make it easier to tighten the bolt? Give a reason for your answer.

(2 marks)

2   Maurice has made a window decoration, as shown in the diagram.
He wants to attach a string to it so that it hangs with the M the right way up.

(a) Maurice's dad says he should find the decoration's centre of mass. How would this help Maurice know where to place the string?

(1 mark)

(b) Describe how Maurice could find the decoration's centre of mass.

(4 marks)

3   (a) Robert says, "A seesaw will only balance if the masses on either side of the pivot are equal." Is Robert correct? Explain your answer.

(2 marks)

(b) A plank AB rests on a pivot 0.4 m from end A. End B is supported by a cord so that the plank is horizontal. The plank is 2.8 m long and weighs 50 N.

    (i)   Draw a diagram of the plank showing the pivot and the forces acting on the plank.

(3 marks)

    (ii) Calculate the tension in the cord.

(2 marks)

Circular Motion

Circular motion — *velocity* is *constantly changing*

1) Velocity is both the speed and direction of an object.

2) If an object is travelling in a circle it is <u>constantly changing direction</u>, which means it's <u>accelerating</u>.

3) This means there <u>must</u> be a <u>force</u> acting on it.

4) This force acts towards the centre of the circle.

5) The force that keeps something moving in a circle is called a <u>centripetal force</u>. *Pronounced sen-tree-pee-tal*

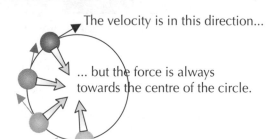

The velocity is in this direction...

... but the force is always towards the centre of the circle.

Tension, *friction* or *gravity* can provide a centripetal force

In the exam, you can be asked to say <u>which force</u> is actually providing the centripetal force in a given situation. It can be <u>tension</u>, or <u>friction</u>, or even <u>gravity</u> (see page 169). Have a look at these examples:

<u>A car going round a bend</u>:

1) Imagine the bend is part of a <u>circle</u> — the centripetal force is towards the <u>centre</u>.

2) The force is from <u>friction</u> between the car's tyres and the road.

friction

tension

<u>A bucket whirling round on a rope</u>:

The centripetal force comes from <u>tension in the rope</u>. Break the rope, and the bucket flies off at a tangent.

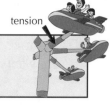

tension

<u>A spinning fairground ride</u>:

The centripetal force comes from <u>tension</u> in the <u>spokes of the ride</u>.

Centripetal force depends on *mass*, *speed* and *radius*

1) The <u>faster</u> an object's moving, the <u>bigger</u> the centripetal force has to be to keep it moving in a <u>circle</u>.

2) And the <u>heavier</u> the object, the <u>bigger</u> the centripetal force has to be to keep it moving in a <u>circle</u>.

3) You also need a <u>larger force</u> to keep something moving in a <u>small circle</u> — it has 'more turning' to do.

Example

Two cars are driving at the same speed around the same circular track. One has a mass of 900 kg, the other has a mass of 1200 kg. Which car has the larger centripetal force?

The <u>three things</u> that mean you need a <u>bigger centripetal force</u> are: <u>more speed</u>, <u>more mass</u>, <u>smaller radius</u> of circle.

In this example, the speed and radius of circle are the same — the <u>only difference</u> is the <u>masses</u> of the cars. So you don't need to calculate anything — you can confidently say:

The <u>1200 kg car</u> (the heavier one) must have the <u>larger centripetal force</u>.

Satellites

A <u>satellite</u> is any object that <u>orbits</u> around a <u>larger object</u> in space. There are natural satellites, like <u>moons</u>, but this page just looks at the artificial ones that we put there ourselves.

*Satellites are **set up** by humans for **many different purposes***

You need to know a few examples of what satellites are used for:

1) Monitoring <u>weather</u> and climate.

2) <u>Communications</u>, e.g. phone and TV.

3) <u>Space research</u>, such as the Hubble Telescope.

4) <u>Spying</u>

5) <u>Navigation</u>, e.g. the Global Positioning System (GPS).

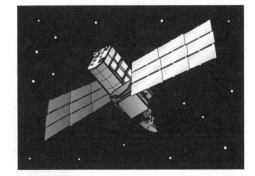

***Communications** satellites **stay** over the **same point** on Earth*

1) Communications satellites are put into <u>quite a high orbit</u> over the <u>equator.</u>

2) This orbit takes <u>exactly 24 hours</u> to complete.

3) This means that the satellites stay above the <u>same point</u> on the Earth's surface, because the Earth <u>rotates below them</u>.

4) So this type of satellite is called a <u>geostationary</u> satellite (geo (= Earth)–stationary) or sometimes a <u>geosynchronous</u> satellite.

5) They're <u>ideal</u> for <u>telephone</u> and <u>TV</u> because they're always in the <u>same place</u> and can <u>transfer signals</u> from one side of the Earth to the other in a <u>fraction of a second</u>.

6) There's room for about <u>400</u> geostationary satellites around the Earth — any more than that and the signals will begin to <u>interfere</u>.

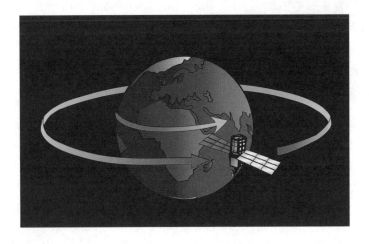

Satellites

So the phone and TV are covered (phew), but the satellites needed for other uses have to move in a <u>different kind of orbit</u> around the Earth.

Weather and spying satellites need to be in a low orbit

1) Geostationary satellites are <u>too high</u> and <u>too stationary</u> to take good <u>weather</u> or <u>spying photos</u> — for this you need <u>low polar orbits</u>, which pass over both <u>poles</u> and are <u>nice and low</u>.

2) In a <u>low polar orbit</u>, the satellite sweeps over <u>both poles</u> whilst the Earth <u>rotates beneath it</u>.

3) The time taken for each full orbit is just <u>a few hours</u>. They're much <u>closer</u> to the Earth than geostationary satellites, so the pull of <u>gravity</u> is stronger and they <u>move much faster</u>.

4) Each time the satellite comes round it can <u>scan</u> the next bit of the globe. This allows the <u>whole surface</u> of the planet to be <u>monitored</u> each day.

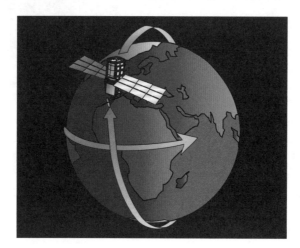

GPS satellites and space telescopes are in other stable orbits

There are satellites in different stable orbits from those above, such as <u>GPS</u> (<u>Global Positioning System</u>) satellites and the <u>Hubble Space Telescope</u>.

But happily you don't need to know anything more than that about them.

Know the two main kinds of orbit used by satellites

In case you're wondering... a GPS system works by transmitting its position and the time. These signals are received by GPS devices in cars or whatever, and once four signals have been received, the device can work out its exact location. It's clever stuff. And you don't need to learn it, which is even better.

Gravity and Orbits

Gravity — it's really the <u>king of the forces</u> in the Universe. It makes <u>stars</u>, it makes things <u>orbit</u> other things, and all the bits that made up the <u>Earth</u> clumped together because of gravity. And it stops us from <u>floating</u> off into space. Good stuff really. Here are some fun gravity facts — enjoy.

Gravity provides the *centripetal force* that causes *orbits*

1) If an object is <u>travelling in a circle</u> it is <u>constantly changing direction</u>, which means there <u>must</u> be a <u>force</u> acting on it (see page 109).

2) An orbit is a <u>balance</u> between the <u>forward motion</u> of the object and a force <u>pulling it inwards</u>. This is a <u>centripetal force</u> — it's directed towards the <u>centre</u> of the circle.

3) The planets move around the Sun in <u>almost circular</u> orbits. The <u>centripetal forces</u> that make this happen are provided by the <u>gravity</u> between each <u>planet</u> and the <u>Sun</u>.

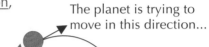

The planet is trying to move in this direction...

...but the force is always towards the centre of the circle.

Gravity decreases quickly as you get *further away*

1) With <u>huge</u> masses like <u>stars</u> and <u>planets</u>, gravity is <u>very strong</u> and acts for <u>a long distance outwards</u>.

2) The <u>closer</u> you get to a star or a planet, the <u>stronger</u> the <u>force of attraction</u>.

3) Because of this stronger force, planets nearer the Sun move <u>faster</u> and cover their orbit more <u>quickly</u>.

4) <u>Moons</u>, <u>artificial satellites</u> and <u>space stations</u> are also held in orbit by gravity. The further out from the Earth their orbit, the slower they move.

5) The size of the force due to gravity follows the fairly famous '<u>inverse square</u>' relationship. The main effect of this is that the force <u>decreases very quickly</u> with increasing <u>distance</u>. The <u>formula</u> is $F \propto 1/d^2$, but I reckon it's <u>easier</u> just to remember the basic idea <u>in words</u>:

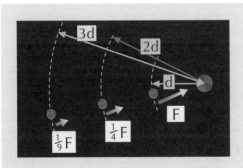

a) If you <u>double the distance</u> from a planet, the size of the <u>force</u> will <u>decrease</u> by a <u>factor of four</u> (2^2).

b) If you <u>treble the distance</u>, the <u>force</u> of gravity will <u>decrease</u> by a <u>factor of nine</u> (3^2), and so on.

c) On the other hand, if you get <u>twice as close</u> the gravity becomes <u>four times stronger</u>.

That's why *comets speed up* and *slow down...*

1) Comets orbit the Sun, but have very <u>eccentric</u> (elongated) orbits with the Sun <u>near one end</u> (see page 87).

2) The comet travels <u>much faster</u> when it's <u>nearer the Sun</u> than it does in the more <u>distant</u> parts of its orbit. That's because the <u>increased pull</u> of gravity makes it <u>speed up</u> the closer it gets to the Sun.

comet

Warm-Up and Exam Questions

Make sure you've learnt this section by doing these questions and the revision summary over the page — then go and have a well-earned break. Come back for Section Seven — you've some real treats in store.

Warm-Up Questions

1) What's the general name for a force that keeps an object moving in a circle?
2) What force keeps satellites in orbit around the Earth?
3) Suggest two uses for satellites.
4) What's the relationship between a body's distance from Earth and the force of gravity on it?
5) At which part of a comet's orbit does it travel most slowly?

Exam Questions

1   A car is driving round a circular track at constant speed.

   (a) Why is the car said to be accelerating when its speed is constant?

   (1 mark)

   (b) (i)   Name two forces acting on the car.

   (1 mark)

        (ii) Are all the forces acting on the car balanced or unbalanced?
              Give a reason for your answer.

   (1 mark)

2   Levi is writing a report on different types of satellite. He is researching two satellites, A and B. Satellite A transmits television signals while B is used to monitor the weather.

   (a) What type of orbit does satellite A use?

   (1 mark)

   (b) How long does satellite A take to complete one orbit?

   (1 mark)

   (c) Satellite B is able to take pictures of the whole surface of the Earth in one day without changing its orbit.

        (i)   Describe how this is possible.

   (2 marks)

        (ii) Which satellite is closer to the Earth, A or B? Give a reason for your answer.

   (1 mark)

3   The Earth is, on average, 1.39 times further from the Sun than Venus.

   (a) Which planet, Earth or Venus, travels faster? Explain your answer.

   (2 marks)

   (b) How many times weaker is the force of the Sun's gravity on Earth than on Venus?

   (2 marks)

Revision Summary for Section Six

Phew. Bit of a mixed bag, that section.

OK, you know what I'm going to say. If you reckon you know your stuff then do these questions and prove it to yourself. If you can't do these questions now, you won't be able to do them in the exam.

1) What's the difference between speed and velocity? Give an example of each.

2)* A tractor travels 2 miles along farm tracks and it takes 15 minutes. It then travels a further 10 miles on a country road and this part of the journey takes 30 minutes.
Calculate the tractor's average speed for the whole journey, in miles per hour.

3)* A boat is sailing due south with a velocity of 0.5 m/s relative to the water. The river is flowing at 0.2 m/s due north. Draw a vector diagram and use it to help you find the boat's resultant velocity.

4)* A bird is facing due north and flying at 12 mph relative to the air. There is a 5 mph wind blowing due west. Draw a vector diagram to help find the resultant velocity of the bird.

5)* Find the distance travelled by a soggy pea as it is flicked from rest to a speed of 14 m/s in 0.4 s. (Assume constant acceleration.)

6) What shape is the trajectory (path) of a projectile?

7)* A sandwich is thrown horizontally off a skyscraper at 1.5 m/s. It hits the ground 10 s later.

   a) How high is the skyscraper? (Take g = 10 m/s². You can ignore air resistance.)

   b) How far will the sandwich have travelled horizontally before it hits the ground?

8) Sarah is levering the lid off a can of paint using a screwdriver. She places the tip of the 20 cm long screwdriver under the can's lid and applies a force of 10 N on the end of the screwdriver's handle as shown. Suggest two ways that Sarah could increase the moment about the pivot point (the side of the can).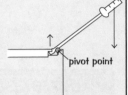

pivot point

9) Describe two different ways of finding the centre of mass of a rectangular playing card.

10)* Arthur weighs 600 N and is sitting on a seesaw 1.5 m from the pivot point. His friend Caroline weighs 450 N and sits on the seesaw so that it balances. How far from the pivot point is she sitting?

11) A cyclist is moving at a constant speed of 7 m/s around a circular track.

   a) Is the cyclist accelerating? Explain your answer.

   b) What force keeps the cyclist travelling in a circle? Where does this force come from?

   c) What will happen to the size of this force if the same cyclist travels at a constant speed of 7 m/s around a different circular track that has a larger radius?

12) State five uses of artificial satellites.

13) Give three differences between a low polar orbit and a geostationary orbit.

14) Gravity is the force of attraction between two masses. What happens to the size of this force if the distance between the masses decreases? Name three important effects of gravity.

15) Two identical satellites orbit at different distances from the Earth. Satellite A orbits the Earth at a distance of 10 000 km and satellite B orbits at 20 000 km. Which satellite has the smaller orbital period? Explain your answer.

16) Explain why comets speed up and slow down during their orbits.

*Answers on page 276.

Magnetic Fields

Electric currents can create magnetic fields. This turns out to be quite useful...

Magnetic fields are areas where a magnetic force acts

Loads of electrical appliances use magnetic fields generated by electric currents.

> A MAGNETIC FIELD is a region where MAGNETIC MATERIALS
> (like iron and steel) and also WIRES CARRYING CURRENTS
> experience a FORCE acting on them.

Magnetic fields can be represented by field diagrams.

The arrows on the field lines always point from the
North pole of the magnet to the South pole.

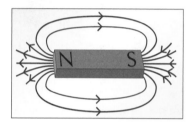

A current-carrying wire creates a magnetic field

There is a magnetic field around a straight, current-carrying wire.

The field is made up of concentric circles with the wire in the centre:

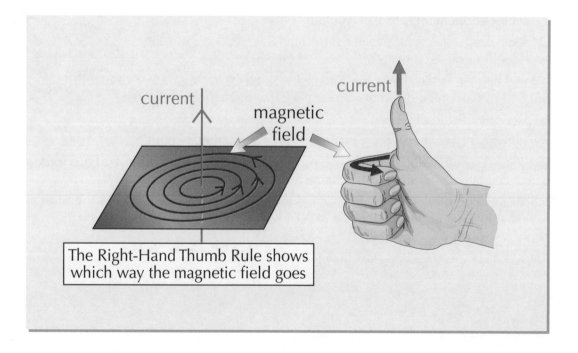

The Right-Hand Thumb Rule shows
which way the magnetic field goes

Just point your thumb in the direction of the current...

...and your fingers show the direction of the field. Remember, it's always your right thumb. Not your
left, but your right thumb. You'll use your left hand on page 175 though, so it shouldn't feel left out...

Magnetic Fields

A **rectangular coil** reinforces the magnetic field

1) If you bend the current-carrying wire round into a <u>coil</u>, the magnetic field looks like this.

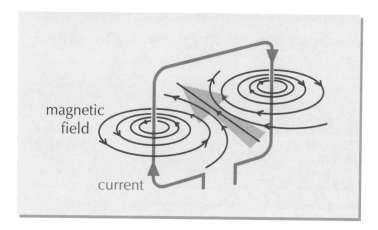

magnetic field

current

2) The circular magnetic fields around the sides of the loop <u>reinforce</u> each other at the centre.

3) If the coil has lots of turns, the fields from all of the individual loops <u>reinforce</u> each other even more.

The magnetic field around a **solenoid**

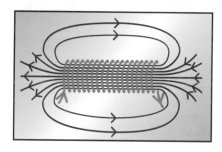

1) The magnetic field <u>inside</u> a current-carrying <u>solenoid</u> (a coil of wire) is <u>strong</u> and <u>uniform</u>.

2) <u>Outside</u> the coil, the field's just like the one round a <u>bar magnet</u>.

3) This means that the <u>ends</u> of a solenoid act like the <u>north pole</u> and <u>south pole</u> of a bar magnet.

4) Pretty obviously, if the <u>direction</u> of the <u>current</u> is <u>reversed</u>, the North and South poles will <u>swap ends</u>.

5) If you imagine looking directly into one end of a solenoid, the <u>direction of current flow</u> tells you whether it's the <u>N or S pole</u> you're looking at, as shown by the <u>diagrams</u> below.

6) You can increase the <u>strength</u> of the magnetic field around a solenoid by adding a magnetically <u>'soft' iron core</u> through the middle of the coil. It's then called an <u>ELECTROMAGNET</u>.

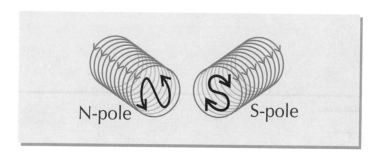

N-pole   S-pole

A <u>magnetically soft</u> material <u>magnetises</u> and <u>demagnetises</u> very easily. So, as soon as you <u>turn off</u> the current through the solenoid, the magnetic field <u>disappears</u> — the iron doesn't stay magnetised.

The Motor Effect

If you put a current-carrying wire into a magnetic field, you have <u>two magnetic fields combining</u>, which puts a force on the wire. The force can make the wire move — which can be quite handy, really.

A **current** in a **magnetic field** experiences a **force**

When a current-carrying wire is put between magnetic poles, the two <u>magnetic fields</u> affect one another. The result is a <u>force</u> on the wire.

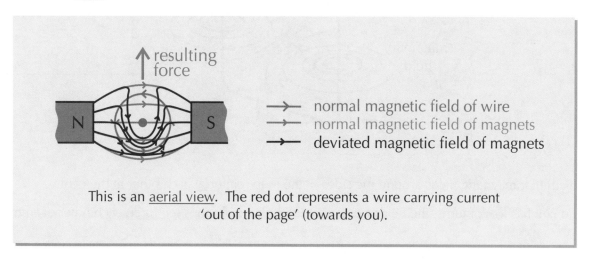

↑ resulting force

→ normal magnetic field of wire
→ normal magnetic field of magnets
→ deviated magnetic field of magnets

This is an <u>aerial view</u>. The red dot represents a wire carrying current 'out of the page' (towards you).

1) To experience the <u>full force</u>, the <u>wire</u> has to be at <u>90°</u> to the <u>magnetic field</u>. If the wire runs <u>along</u> the <u>magnetic field</u>, it won't experience <u>any force at all</u>. At angles in between, it'll feel <u>some</u> force.

2) The <u>force</u> gets <u>bigger</u> if either the <u>current</u> or the <u>magnetic field</u> is made bigger.

3) The direction the force acts in depends on the <u>direction</u> of the <u>magnetic field</u> and the <u>direction of the current</u> in the wire — as shown in the diagram on the right.

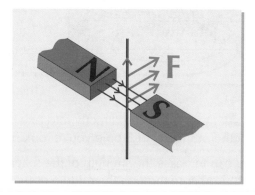

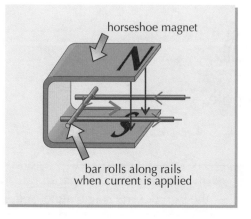

horseshoe magnet

bar rolls along rails when current is applied

4) A good way of showing the <u>direction of the force</u> is to apply a current to a set of rails inside a <u>horseshoe magnet</u>. A <u>bar</u> is placed on the rails, which <u>completes the circuit</u>. This generates a <u>force</u> that <u>rolls the bar</u> along the rails.

The Motor Effect

Fleming's left-hand rule *tells you* **which way** *the force acts*

1) They could test to see if you can do this, so practise it.

2) Using your left hand, point your **F**irst finger in the direction of the **F**ield and your se**C**ond finger in the direction of the **C**urrent.

3) Your thu**M**b will then point in the direction of the force (**M**otion).

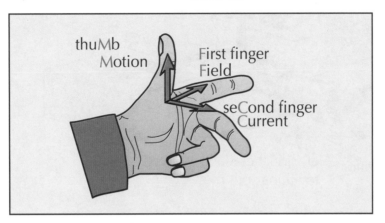

Example:

Which direction is the force on the wire?

Answer:

1) Draw in arrows for current (positive to negative) and magnetic field (north to south):

2) Line up your fingers with the arrows using Fleming's LHR.

3) Draw in direction of force (motion) according to which way your thumb's pointing:

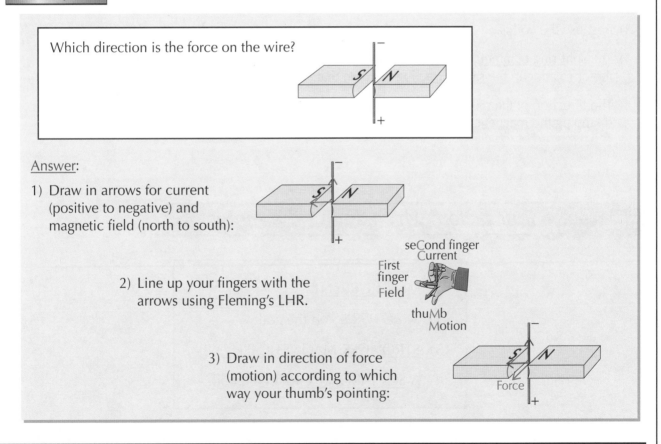

First finger = Field, seCond finger = Current, thuMb = Motion

See, I told you you'd need your left hand for this page. You might think thu**M**b = **M**otion is a bit of a tenuous link, but I bet you never forget it now. It's a bit like when you really hate an advert on TV. Anyway, learn the rule and use it — don't be scared of looking like a muppet in the exam.

The Simple Electric Motor

Aha — one of the favourite exam topics of all time. Read it. Understand it. Learn it.

The simple **electric motor**

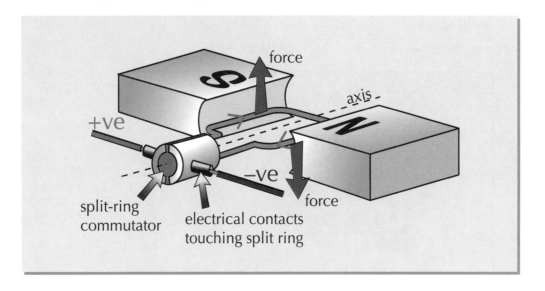

1) The diagram shows the <u>forces</u> acting on the two <u>side arms</u> of the <u>coil</u>.

2) These forces are just the <u>usual forces</u> which act on <u>any current</u> in a <u>magnetic field</u>.

3) Because the coil is on a <u>spindle</u> and the forces act <u>one up</u> and <u>one down</u>, it <u>rotates</u>.

4) The <u>split-ring commutator</u> is a clever way of '<u>swapping</u> the contacts <u>every half turn</u> to keep the motor rotating in the <u>same direction</u>'. (Learn that statement because they might ask you about it.)

5) The direction of the motor can be <u>reversed</u> either by swapping the <u>polarity</u> of the <u>DC supply</u> or swapping the <u>magnetic poles</u> over.

There are **four** factors which **speed the motor up**

> 1) More <u>CURRENT</u>.
>
> 2) More <u>TURNS</u> on the coil.
>
> 3) <u>STRONGER MAGNETIC FIELD</u>.
>
> 4) A <u>SOFT IRON CORE</u> in the coil.

For example...

Loudspeakers demonstrate the <u>motor effect</u>. <u>AC electrical signals</u> from the <u>amplifier</u> are fed to the <u>speaker coil</u> (shown in red). These make the coil move <u>back and forth</u> over the poles of the <u>magnet</u>. These movements make the <u>cardboard cone vibrate</u> and this creates <u>sounds</u>.

The Simple Electric Motor

Answering <u>questions</u> on these might often involve using what you've learnt on earlier pages:

Example:

> Is the coil turning clockwise or anticlockwise?
>
>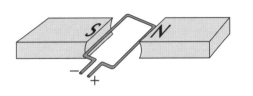

<u>Answer:</u>

1) Draw in current arrows (positive to negative):

   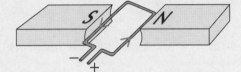

2) Fleming's LHR on one arm (I've used the right-hand arm).

   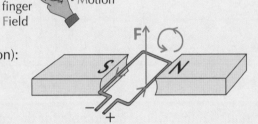

   SeCond finger — Current
   First finger — Field
   thuMb — Motion

3) Draw in direction of force (motion):

   So — the coil is turning <u>anticlockwise</u>.

Practical motors *have pole pieces which are* very curved

1) Link the coil to an <u>axle</u>, and the axle <u>spins round</u>.

2) If you can make your motor powerful enough, that axle can turn just about anything.

3) The problem is that the type of motor shown in the diagram on the last page is pretty useless. It's too <u>inefficient</u> to power anything big and heavy.

4) Instead, practical motors use <u>pole pieces</u> which are <u>so curved</u> that they form a <u>hollow cylinder</u>. The coil spins inside the cylinder.

curved pole pieces of magnet

N — S

coil

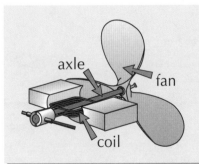

axle

fan

coil

In this diagram there's a <u>fan</u> attached to the axle, but you can stick <u>almost anything</u> on a motor axle and make it spin round. For example, in a <u>food mixer</u> the axle's attached to a <u>blade</u> or whisks. In a <u>CD player</u> the axle's attached to the bit you <u>sit the CD on</u>.

Warm-Up and Exam Questions

It's time for some questions — make sure you can do the warm-up ones first, then have a go at the exam questions below.

Warm-Up Questions

1) What does the right-hand thumb rule show?
2) Draw a diagram to show why a current-carrying wire in a magnetic field experiences a force.
3) In Fleming's left-hand rule, what's represented by the first finger? the second finger? the thumb?
4) Suggest two uses for electromagnets.
5) What are the four factors that affect the speed of an electric motor?

Exam Questions

1   Arnold is making an electromagnet using a current-carrying solenoid and a core.

   (a) Complete the following diagram of the solenoid to show the magnetic field around it.

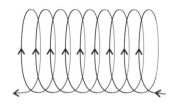

(2 marks)

   (b) Suggest a material that would be suitable for the core.

(1 mark)

   (c) The core material Arnold has chosen is magnetically soft. Explain what this means.

(1 mark)

   (d) What effect will this core have on the magnetic field produced by the electromagnet?

(1 mark)

2   Julia is designing a toy car with a small electric motor to drive the wheels.

The diagram shows a simplified version of Julia's motor.

   (a) In which direction will the wheel turn?

(1 mark)

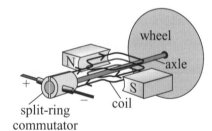

   (b) Explain the purpose of the split-ring commutator.

(1 mark)

   (c) Julia is testing her car and wants it to go faster.
Suggest a change Julia can make to the motor to make the car go faster.

(1 mark)

   (d) Julia wants to make the car drive in forward and reverse.
How can she reverse the direction of the motor?

(1 mark)

Generators

Think about the simple electric <u>motor</u> — you've got a current in the wire and a magnetic field, which causes movement. Well, a <u>generator</u> works the <u>opposite way round</u> — you've got a magnetic field and movement, which <u>induces a current</u>.

AC generators — *just turn the* **coil** *and there's a current*

You've already met <u>generators</u> and <u>electromagnetic induction</u> (see page 40) — this is a bit more detail about how a simple generator works.

1) Generators <u>rotate a coil</u> in a <u>magnetic field</u> (or a magnet in a coil... see below).

2) Their <u>construction</u> is much like a <u>motor</u>.

3) As the <u>coil spins</u>, a <u>current</u> is <u>induced</u> in the coil. This current <u>changes direction</u> every half turn.

4) Instead of a <u>split-ring commutator</u>, AC generators have <u>slip rings</u> and <u>brushes</u> so the contacts <u>don't swap</u> every half turn.

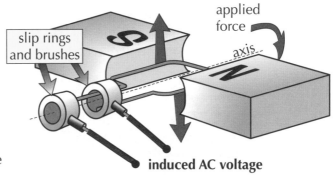

induced AC voltage

original   faster revs

5) This means they produce <u>AC voltage</u>, as shown by these <u>CRO displays</u>. Note that <u>faster revolutions</u> produce not only <u>more peaks</u> but <u>higher overall voltage</u> too.

6) All power stations use AC generators to produce electricity, they just get the energy needed to turn the coil or magnetic field in different ways.

Dynamos — *you* **turn** *the* **magnet** *instead of the coil*

1) <u>Dynamos</u> are a slightly different type of <u>generator</u>. They rotate the <u>magnet</u> instead of the coil.

2) This still causes the <u>field through the coil</u> to <u>swap</u> every half turn, so the output is <u>just the same</u> as for a generator.

3) This means you get the <u>same CRO traces</u> of course.

1) <u>Dynamos</u> are sometimes used on <u>bikes</u> to power the <u>lights</u>.

2) The <u>cog wheel</u> at the top is positioned so that it <u>touches</u> one of the <u>bike wheels</u>.

3) As the wheel moves round, it <u>turns</u> the cog which is attached to the <u>magnet</u>.

4) This creates an <u>AC current</u> to power the lights.

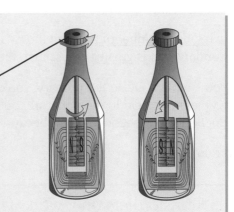

Transformers

Transformers use electromagnetic induction. So they will only work on AC.

Transformers *change the* **voltage** *— but only* **AC voltages**

There are a few different types of transformer. The ones you need to know are step-up transformers, step-down transformers and isolating transformers (see page 182). They all have two coils, the primary and the secondary, joined with an iron core.

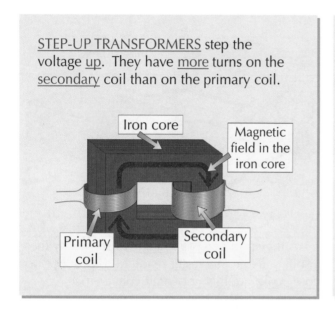

STEP-UP TRANSFORMERS step the voltage up. They have more turns on the secondary coil than on the primary coil.

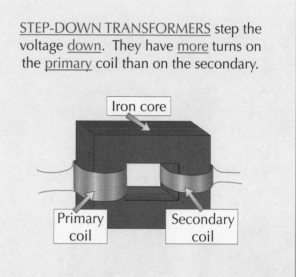

STEP-DOWN TRANSFORMERS step the voltage down. They have more turns on the primary coil than on the secondary.

Transformers *work by* **electromagnetic induction**

1) The primary coil produces a magnetic field which stays within the iron core. This means nearly all of it passes through the secondary coil and hardly any is lost.

2) Because there is alternating current (AC) in the primary coil, the field in the iron core is constantly changing direction (100 times a second if it's at 50 Hz) — i.e. it is a changing magnetic field.

3) This rapidly changing magnetic field is then felt by the secondary coil.

4) The changing field induces an alternating voltage in the secondary coil (with the same frequency as the alternating current in the primary) — electromagnetic induction of a voltage in fact.

5) The relative number of turns on the two coils determines whether the voltage induced in the secondary coil is greater or less than the voltage in the primary.

6) If you supplied DC to the primary, you'd get nothing out of the secondary at all. Sure, there'd still be a magnetic field in the iron core, but it wouldn't be constantly changing, so there'd be no induction in the secondary coil because you need a changing field to induce a voltage.

So don't forget it —

Transformers only work with AC. They don't work with DC at all.

Transformers

The *iron core* carries *magnetic field, not current*

1) The <u>iron core</u> is purely for transferring the <u>changing magnetic field</u> from the primary coil to the secondary.

2) No <u>electricity</u> flows round the <u>iron core</u>.

3) The iron core can be <u>laminated</u> with <u>layers of insulation</u> to reduce <u>eddy currents</u> in the iron. Eddy currents are little 'whirlpools' of charge that build up in the iron, <u>heating it up</u> and <u>wasting energy</u>.

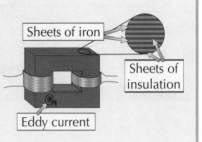

Sheets of iron

Sheets of insulation

Eddy current

Transformers are nearly *100% efficient* so *'power in = power out'*

The formula for <u>power supplied</u> is: <u>power = voltage × current</u> or: <u>$P = V \times I$</u>.

So you can rewrite <u>power in = power out</u> as:

$$V_p I_p = V_s I_s$$

V_p = primary voltage	V_s = secondary voltage
I_p = primary current	I_s = secondary current

The *transformer equation* — use it *either way up*

You can calculate the output voltage from a transformer if you know the input voltage and the number of turns on each coil.

$$\frac{\text{primary voltage}}{\text{secondary voltage}} = \frac{\text{number of turns on primary}}{\text{number of turns on secondary}}$$

$$\frac{V_P}{V_S} = \frac{N_P}{N_S}$$

or

$$\frac{V_S}{V_P} = \frac{N_S}{N_P}$$

Well, it's <u>just another formula</u>. You stick in the numbers <u>you've got</u> and work out the one <u>that's left</u>. It's really useful to remember you can write it <u>either way up</u> — this example's much trickier algebra-wise if you start with V_s on the bottom...

Example:

A transformer has 40 turns on the primary and 800 on the secondary. If the input voltage is 1000 V, find the output voltage.

<u>Answer:</u> $\frac{V_s}{V_p} = \frac{N_s}{N_p}$, so $\frac{V_s}{1000} = \frac{800}{40}$

$V_s = 1000 \times \frac{800}{40} = 20\ 000$ V

Or you can say that 800 is 20 times 40, so the secondary voltage will also be 20 times the primary voltage.

Transformers

Transformers are used on the National Grid

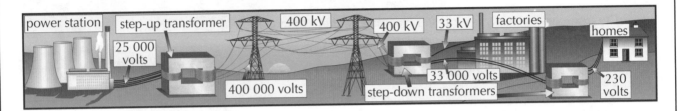

You get both step-up and step-down transformers on the National Grid:

1) To transmit <u>a lot of power</u>, you either need <u>high voltage</u> or <u>high current</u> (P = VI).

2) The problem with <u>high current</u> is the <u>loss</u> (as heat) due to the <u>resistance</u> of the cables.

3) The formula for <u>power loss</u> due to resistance in the cables is: $\underline{P = I^2 R}$.

4) Because of the I^2 bit, if the current is <u>10 times</u> bigger, the losses will be <u>100 times</u> bigger.

5) It's much <u>cheaper</u> to boost the voltage up to <u>400 000 V</u> and keep the current <u>very low</u>.

6) This requires <u>transformers</u> as well as <u>big pylons</u> with <u>huge insulators</u>, but it's still <u>cheaper</u>.

7) The transformers have to <u>step</u> the voltage <u>up</u> at one end, for <u>efficient transmission</u>, and then bring it back down to <u>safe, useable levels</u> at the other end.

Isolating transformers are used in bathrooms

1) Most household transformers <u>reduce</u> the mains voltage for use in <u>low-voltage</u> devices such as radios.

2) However, <u>isolating</u> transformers have <u>equal</u> primary and secondary voltages. That means they have <u>equal numbers of turns</u> on the primary and secondary coils. This is because the only purpose of an isolating transformer is <u>safety</u>.

3) The <u>danger</u> of the <u>mains</u> circuit is that it's connected to the <u>earth</u>, so if you <u>touch</u> the <u>live parts</u> and are <u>also touching the ground</u>, you will <u>complete a circuit</u> with you in it. <u>NOT good</u>.

4) The isolating transformer inside a bathroom shaver socket allows you to use the shaver without being <u>physically connected</u> to the mains. So it minimises the risk of the <u>live</u> parts <u>touching</u> the <u>earth</u> lead and likewise <u>minimises your risk</u> of getting <u>electrocuted</u>.

A very high voltage is used in the cables and a lower one in your home

In most power stations, <u>fuels</u> are burned (or uranium is split) and the energy from this powers a huge <u>generator</u>. Not all the heat can be converted into mechanical power, though, so heat is often <u>lost</u> to the environment. If the plant <u>is</u> able to reuse the heat, it's referred to as a <u>cogeneration power plant</u>.

Warm-Up and Exam Questions

It's that time again — check you can do the basics, then get stuck into some lovely exam questions. Don't forget to go back and check up on any niggling bits you can't do.

Warm-Up Questions

1) AC generators do not have a split-ring commutator. What do they have instead?
2) Write down the transformer equation.
3) What are isolating transformers and what are they used for?
4) Why is electricity distributed at high voltage?

Exam Questions

1 Explain how a generator produces an alternating current.

(3 marks)

2 Gordon is thinking of fitting a dynamo to his bicycle, to power its lights. He is investigating the voltage that might be produced when he cycles by connecting a dynamo to an oscilloscope.

(a) Match the most appropriate CRO trace shown to the following conditions:

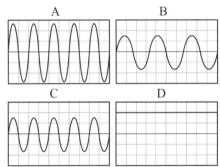

    (i) The dynamo is turned slowly.

(1 mark)

    (ii) The dynamo is turned quickly.

(1 mark)

(b) (i) What happens to the dynamo's output when it is not rotating?

(1 mark)

    (ii) Why is this a disadvantage when a dynamo is used to power bicycle lights?

(1 mark)

3 A student is trying to test a transformer using a battery.

(a) Explain why the voltmeter connected to the secondary coil reads 0 V.

(2 marks)

(b) The student finds an AC power supply and reconnects the transformer. Her results are: $V_P = 12$ V, $I_P = 2.5$ A, $V_S = 4$ V (where V_P is the voltage across the primary coil, etc.).

    (i) Calculate the power input to the transformer.

(1 mark)

    (ii) Calculate the current in the secondary coil, I_S.

(2 marks)

    (iii) The primary coil has 15 turns. How many turns must be on the secondary coil?

(2 marks)

Nuclear Power

So, generators to... well... generate electricity, transformers to help distribute it, it all sounds just lovely. Most power stations get the energy they need to drive the generators by <u>burning fuel</u> (e.g. coal) or from the <u>natural motion</u> of something (e.g. waves, tides). <u>Nuclear</u> power stations do it a bit differently...

Nuclear power stations are really glorified steam engines

Nuclear power stations are powered by <u>nuclear reactors</u>. In a nuclear reactor, a <u>controlled chain reaction</u> takes place in which uranium or plutonium atoms <u>split up</u>, releasing <u>loads</u> of <u>energy</u> in the form of <u>heat</u>.

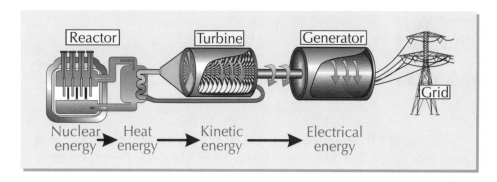

You can split more than one atom — fission chain reactions

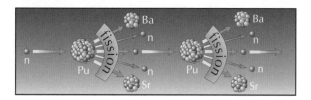
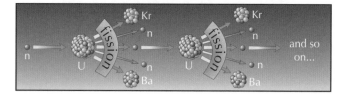

1) The 'fuel' that's split in a nuclear reactor (and in <u>bombs</u> — a nuclear bomb is an <u>uncontrolled</u> fission reaction) is usually either <u>uranium-235</u> or <u>plutonium-239</u>.

2) If a <u>slow-moving neutron</u> gets absorbed by a uranium or plutonium nucleus, the nucleus can <u>split</u>.

3) Each time a <u>uranium</u> or <u>plutonium</u> nucleus <u>splits up</u>, it spits out <u>two or three neutrons</u>. These might go on to hit other nuclei, causing them to split and release even more neutrons, which hit even more nuclei... and so on and so on. This process is known as a chain reaction.

4) When a large atom splits in two it will form <u>two new lighter elements</u>. These new nuclei are usually <u>radioactive</u> because they have the 'wrong' number of neutrons in them.

5) This is the <u>big problem</u> with nuclear power — it produces <u>huge</u> amounts of <u>radioactive material</u> which is very <u>difficult</u> and <u>expensive</u> to dispose of safely.

6) Each nucleus <u>splitting</u> (called a <u>fission</u>) gives out <u>a lot of energy</u> — a lot more energy than you get from a <u>chemical</u> bond between two atoms (e.g. by burning).

Nuclear power releases a lot of energy, but it has its downsides...

Nothing to it really, chuck in some sluggish neutrons, split some atoms, get some heat, make some steam, turn a turbine, drive a generator and ta da — some <u>electricity</u>. Then you have to deal with the nasty <u>radioactive mess</u> that's left behind, which can be a bit of a worry (see the next page).

Nuclear Power

Inside a gas-cooled *nuclear reactor*

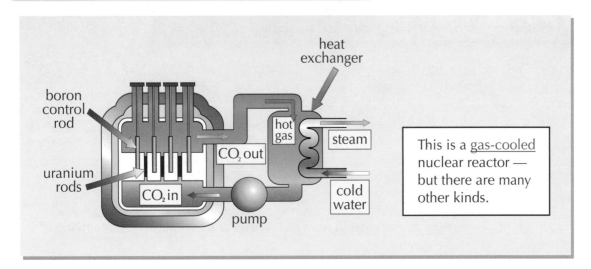

This is a gas-cooled nuclear reactor — but there are many other kinds.

1) <u>Free neutrons</u> in the reactor '<u>kick-start</u>' the fission process.

2) The atoms <u>produced</u> then <u>collide</u> with other atoms, causing the <u>temperature</u> in the reactor to <u>rise</u>.

3) <u>Control rods</u>, often made of <u>boron</u>, limit the rate of fission by <u>absorbing</u> excess neutrons.

4) A gas, typically <u>carbon dioxide</u>, is pumped through the <u>reactor</u> to carry away the <u>heat</u> generated.

5) The gas is then passed through a <u>heat exchanger</u>, where it gives its energy to <u>water</u> — this water is heated and turned into <u>steam</u>, which turns a <u>turbine</u>, generating electricity.

Using *nuclear power* has its *pros* and *cons*

1) <u>Fossil fuels</u> (coal, oil and gas) all release CO_2 when they're burnt. This adds to the <u>greenhouse effect</u> and <u>global warming</u>. Burning coal and oil also releases <u>sulfur compounds</u> that can cause <u>acid rain</u>.

2) In terms of emissions like these, <u>nuclear power</u> is very very <u>clean</u>.

3) The <u>main environmental problem</u> is with the disposal of <u>waste</u>. The products left over after nuclear fission are generally <u>radioactive</u>, so they can't just be thrown away.

4) There are a few ways to dispose of the waste. One way is to pack the waste into <u>thick metal containers</u>, put the containers into a <u>very deep hole</u>, then fill the hole with <u>concrete</u>.

5) Some of this waste will stay dangerously radioactive for <u>hundreds of years</u>, and some people worry that materials could <u>leak out</u> of the storage facilities over time.

6) And <u>nuclear power</u> always carries the risk of <u>leaks</u> from the plant or of a <u>major catastrophe</u> like <u>Chernobyl</u> (where there was a nuclear explosion at a plant).

7) But it's not all doom and gloom. Building a nuclear plant can have a very <u>positive impact</u> on an area. Both the plant itself, and the <u>support industries</u> that spring up around it, bring lots of <u>skilled jobs</u> to rural areas like the west coast of Cumbria.

8) Nuclear <u>fuel</u> (i.e. the uranium) is <u>cheap</u> but the <u>overall cost</u> of nuclear power is <u>high</u> due to the cost of the <u>power plant</u> and final <u>decommissioning</u>. Dismantling a nuclear plant safely takes <u>decades</u>.

Nuclear Fusion

The main problem with nuclear fission is that there's lots of radioactive mess to clean up afterwards. This is why scientists have been looking into producing energy the same way stars do — through <u>fusion</u>.

Nuclear fusion — the *joining* of small atomic nuclei

1) <u>Nuclear fusion</u> is just the <u>opposite</u> of fission — two <u>light nuclei</u> (e.g. hydrogen) can <u>combine</u> to create a larger nucleus.

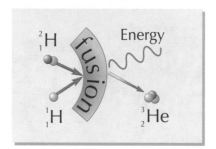

2) Fusion releases <u>a lot</u> of energy (more than fission for a given mass of fuel). So people are trying to develop <u>fusion reactors</u> to make <u>electricity</u>.

3) Fusion <u>doesn't</u> leave behind a lot of radioactive <u>waste</u> and there's <u>plenty</u> of hydrogen about for <u>fuel</u>.

4) The <u>big problem</u> is that fusion only happens at <u>really high temperatures</u> — over <u>10 000 000 °C</u>.

5) <u>No material</u> can stand that kind of temperature without being <u>vaporised</u>, so fusion reactors are really difficult to build. You have to contain the hot hydrogen in a <u>magnetic field</u> instead of in a physical container.

6) There are a few <u>experimental</u> reactors around, but none of them are generating electricity yet. At the moment it takes <u>more</u> power to get up to the right temperature than the reactor can produce.

Cold fusion — hoax or energy of the future?

1) Cold fusion is <u>nuclear fusion</u> which occurs at around <u>room temperature</u>.

2) In 1989 two scientists, <u>Stanley Pons</u> and <u>Martin Fleischmann</u>, claimed to have released energy from cold fusion using a simple experiment.

3) This caused a lot of <u>excitement</u> — cold fusion would make it possible to generate lots of electricity, easily and cheaply.

4) Many scientists were <u>sceptical</u>, and the results have never been repeated reliably enough to be accepted by the scientific community.

5) Not all scientists have given up though — there's still a lot of research into cold fusion, so you never know...

And as usual, some people see only the potential for a great big bomb

At about the same time as research started on fusion reactors, physicists were working on a <u>fusion bomb</u>. These 'hydrogen bombs' are incredibly powerful — they can release a few thousand times more energy than the nuclear bombs that destroyed Hiroshima and Nagasaki at the end of World War II.

Warm-Up and Exam Questions

The end of another section — they just go far too quickly. Make sure you've understood it all by doing these questions (and the revision summary on the next page) before you whizz on to Section Eight.

Warm-Up Questions

1) State the energy transfers that occur when electricity is produced from nuclear energy.
2) Name two elements often used as nuclear fuel.
3) What does nuclear fission produce in addition to energy? Why is this a problem?
4) Why were people excited about Pons and Fleischmann's cold fusion experiment?
5) Why have Pons and Fleischmann's results not been accepted by the scientific community?

Exam Questions

1   Sparktown Electricity Company is planning a new nuclear power station. They are holding a public meeting with people who live in the area to discuss the power station.

(a) Give two worries that people in the area might have about the planned power station.

(2 marks)

(b) Suggest two benefits a nuclear power station might have for the area.

(2 marks)

(c) The spokesperson from the electricity company says that nuclear power is more environmentally friendly than fossil fuels. Is this true? Give a reason for your answer.

(1 mark)

(d) A man at the meeting asks, "I understand that nuclear power stations and nuclear bombs both use chain reactions — doesn't this mean nuclear power is very dangerous?"

(i)   Describe the chain reaction used in nuclear power plants.

(3 marks)

(ii) How are the chain reactions in a power station made safer than those in a bomb?

(2 marks)

2   The table shows some information about various elements and isotopes.

Element	Deuterium	Hydrogen	Krypton	Plutonium	Thorium	Tin
Relative atomic mass	2	1	84	239	232	119

(a) Which two substances in the table would be most likely to be used in a fusion reaction?

(2 marks)

(b) Explain why scientists are interested in developing fusion power.

(2 marks)

(c) Why is fusion not used to generate electricity at present?

(1 mark)

Revision Summary for Section Seven

Electricity and magnetism. What joys. And a bit of fission and fusion thrown in just for fun. This is definitely physics at its most grisly. The big problem with physics in general is that usually there's nothing to 'see'. You're told that there's a current flowing or a magnetic field lurking, but there's nothing you can actually see with your eyes. That's what makes it so difficult, but it's what you need to get used to in physics I'm afraid. Ah well, do these questions to see how you're getting on.

1) Give a definition of a magnetic field.
2) Sketch magnetic fields for: a) a current-carrying wire, b) a rectangular coil, c) a solenoid.
3) What is an electromagnet made of?
4) Explain how to work out the polarity of the ends of an electromagnet.
5) Explain what is meant by 'magnetically soft'.
6) Explain what Fleming's Left-Hand Rule is for and how it works.
7) Sketch a simple electric motor and list the four ways to speed it up.
8) What is electromagnetic induction? List four factors which affect the size of the induced voltage.
9) Sketch a generator, labelling all the parts. Describe how it works and what all the bits do.
10) What is a dynamo?
11) Give an example of how a dynamo can be used.
12) Sketch a step-up and a step-down transformer and explain how they work.
13)*In a transformer, the primary voltage is 6 V, the primary current is 10 A and the secondary voltage is 3 V. What is the secondary current in the transformer?
14)*A transformer has an input voltage of 20 V and an output voltage of 16 V.
     If there are 64 turns on the secondary coil, how many turns are there on the primary coil?
15) Make a sketch of how transformers are used in the National Grid.
16) Explain why power is transmitted at such a high voltage.
17) Write down three facts about isolating transformers.
18) Explain in terms of energy transfers how electricity is produced in a nuclear power station.
19) What is nuclear fission?
20) Explain how the chain reaction in a nuclear reactor works.
21) What is used in a nuclear reactor to absorb excess neutrons?
22) Give one reason for using nuclear power rather than fossil fuels.
23) What is the main environmental problem associated with nuclear power?
24) What is nuclear fusion? Why is it difficult to construct a working fusion reactor?
25) What is cold fusion?
26) Go and read up on quantum theory... no wait, I mean... go and put the kettle on.

* Answers on page 277.

Images

Mirror, mirror on the wall... yes, you may be beautiful, but that won't help you in your exams. Knowing about <u>reflection</u> and <u>refraction</u> will. But brace yourself — there are a <u>lot</u> of diagrams on the next pages.

A **real** image is **actually there** — a **virtual** image is **not**

1) A <u>real image</u> is where the <u>light from an object</u> comes together to form an <u>image on a 'screen'</u> — like the image formed on an eye's <u>retina</u> (the 'screen' at the back of an <u>eye</u>).

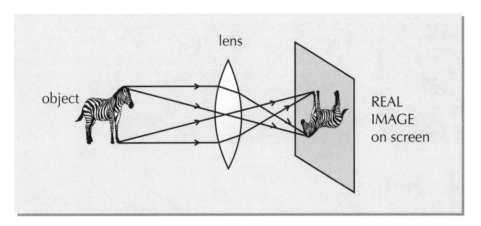

2) A <u>virtual image</u> is when the rays are diverging, so the light from the object <u>appears</u> to be coming from a completely <u>different place</u>.

3) When you look in a <u>mirror</u> you see a <u>virtual image</u> of your face — because the <u>object</u> (your face) <u>appears</u> to be <u>behind the mirror</u>.

4) You can get a virtual image when looking at an object through a <u>magnifying lens</u> — the virtual image looks <u>bigger</u> and <u>further away</u> than the object <u>actually</u> is.

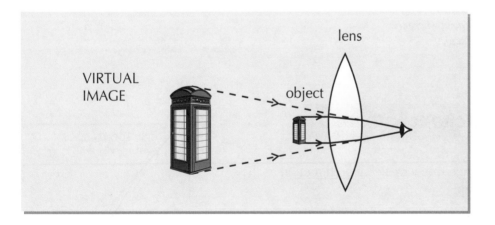

To describe an image properly, you need to say <u>three things</u>:

> 1) <u>How big it is</u> compared to the object
>
> 2) Whether it's <u>upright or inverted</u> (upside down)
>
> 3) Whether it's <u>real or virtual</u>.

Images

Reflection of light lets you see things

1) <u>Reflection of light</u> is what allows us to <u>see</u> objects. Light bounces off them into our eyes.

2) When light reflects off an <u>uneven surface</u> such as a <u>piece of paper</u>, the light reflects off <u>at all different angles</u> and you get a <u>diffuse reflection</u>.

3) When light reflects from an <u>even surface</u> (something <u>smooth and shiny</u> like a <u>mirror</u>) then it's all reflected at the <u>same angle</u> and you get a <u>clear reflection</u>.

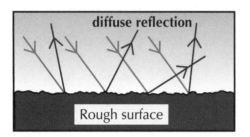

diffuse reflection

Rough surface

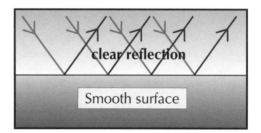

clear reflection

Smooth surface

4) But don't forget, the <u>LAW OF REFLECTION</u> applies to <u>every reflected ray</u>:

angle of <u>incidence</u> = angle of <u>reflection</u>

Note that these two angles are <u>ALWAYS</u> defined between the ray itself and the <u>dotted NORMAL</u>. <u>Don't ever</u> label them as the angle between the ray and the <u>surface</u>.

The <u>normal</u> is an imaginary line that's at right angles to the surface (at the point where the light hits the surface).

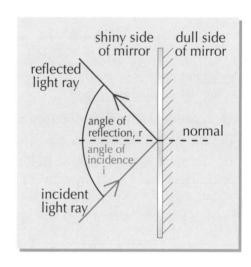

shiny side of mirror — dull side of mirror

reflected light ray

angle of reflection, r

normal

angle of incidence, i

incident light ray

Light bends as it changes speed

1) <u>Refraction</u> of light is when the waves <u>change direction</u> as they <u>enter a different medium</u>.

2) This is caused <u>entirely</u> by the <u>change in speed</u> of the waves.

3) That's what makes ponds <u>look shallower</u> than they are — light reflects off the <u>bottom</u> and <u>speeds up</u> when it <u>leaves the water</u>, making the bottom look like it's <u>nearer</u> than it is:

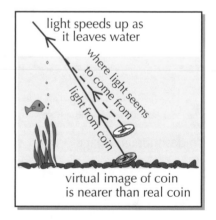

light speeds up as it leaves water

where light seems to come from

light from coin

virtual image of coin is nearer than real coin

Take your time — this stuff about images can seem really odd at first

Make sure you've learnt this little lot well enough to answer typical exam questions like these:
<u>"Explain why you can see a piece of paper."</u>   <u>"What is diffuse reflection?"</u>
<u>"What is a clear reflection?"</u>   <u>"Why do light rays bend as they leave a pond?"</u>

Mirrors

The examiners do like to see a nice diagram, so get your rulers out.

Draw a *ray diagram* for an *image* in a *plane mirror*

You need to be able to <u>reproduce</u> this entire diagram of <u>how an image is formed</u> in a <u>PLANE MIRROR</u>.

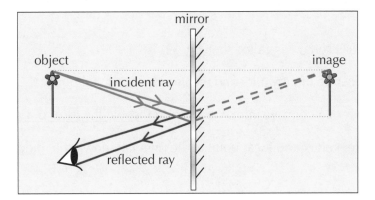

Learn these <u>three important points</u>:

> 1) The <u>image</u> is the <u>same size</u> as the <u>object</u>.
>
> 2) It is <u>AS FAR BEHIND</u> the mirror as the object is <u>in front</u>.
>
> 3) It's formed from <u>diverging rays</u>, which means it's a <u>virtual image</u>.

1) First off, draw the <u>virtual image</u>. <u>Don't</u> try to draw the rays first. Follow the rules in the above box — the image is the <u>same size</u>, and it's <u>as far behind</u> the mirror as the object is in <u>front</u>.

2) Next, draw a <u>reflected ray</u> going from the top of the virtual image to the top of the eye. Draw a <u>bold line</u> for the part of the ray between the mirror and eye, and a <u>dotted line</u> for the part of the ray between the mirror and the virtual image.

3) Now draw the <u>incident ray</u> going from the top of the object to the mirror. The incident and reflected rays follow the <u>law of reflection</u> — but you <u>don't</u> actually have to measure any angles. Just draw the ray from the <u>object</u> to the <u>point</u> where the reflected ray <u>meets the mirror</u>.

4) Now you have an <u>incident ray</u> and <u>reflected ray</u> for the <u>top</u> of the image. Do <u>steps 2 and 3 again</u> for the <u>bottom</u> of the <u>eye</u> — a reflected ray going from the image to the bottom of the eye, then an incident ray from the object to the mirror.

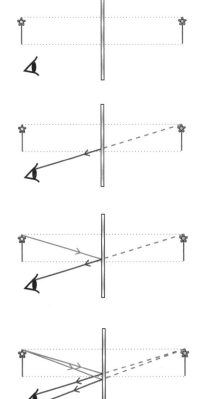

Mirrors

Curved mirrors are a little more complicated

Concave mirrors are shiny on the inside of the curve and convex mirrors are shiny on the outside. Light shining on a concave mirror converges, and light shining on a convex mirror diverges.

1) Uniformly curved mirrors are like a round portion of a sphere. The centre of the sphere is the centre of curvature, C.

2) The centre of the mirror's surface is called the vertex.

3) Halfway between the centre of curvature and the vertex is the focal point, F.

4) Rays parallel to the axis of a concave mirror reflect and meet at the focal point.

5) The centre of curvature, vertex and focal point all lie on a line down the middle of the mirror called the axis.

6) The centre of curvature and focal point are in front of a concave mirror and behind a convex mirror.

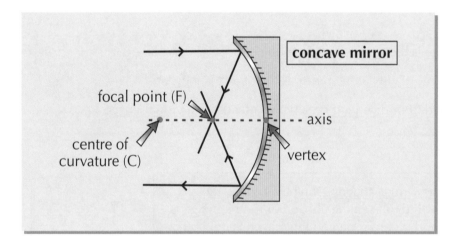

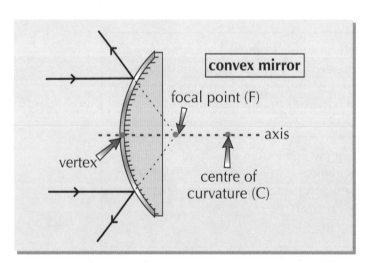

You must be able to label F, C and the vertex on a curved mirror

Reflection in a plane (flat) mirror is not too hard to learn. Go through the method step by step — doing it "image first" instead of "rays first" gives you a nice neat diagram, even if it seems weird to do it that way round. Learn all the facts about curved mirrors — it'll make the next pages a load easier.

Mirrors

You could also get asked to draw a <u>ray diagram</u> of reflection in a curved mirror. Pay attention, it's tricky.

*Draw a **ray diagram** for an **image** in a **concave mirror***

1) An incident ray <u>parallel to the axis</u> will pass through the <u>focal point</u> when it's reflected.
2) An incident ray passing <u>through the focal point</u> will be <u>parallel to the axis</u> when it's reflected.

1) Pick a point on the <u>top</u> of the object. Draw a ray going from the object to the mirror <u>parallel</u> to the axis of the mirror.

2) Draw another line going from the top of the <u>object to the mirror</u>, passing through the <u>focal point</u> on the way.

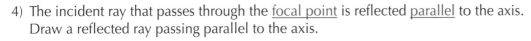

3) The incident ray that's <u>parallel</u> to the axis is <u>reflected</u> through the <u>focal point</u>. Draw a <u>reflected ray</u> passing through the focal point.

4) The incident ray that passes through the <u>focal point</u> is reflected <u>parallel</u> to the axis. Draw a reflected ray passing parallel to the axis.

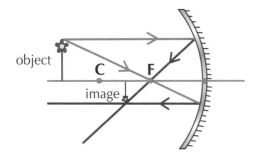

5) Mark where the two reflected rays <u>meet</u>. That's the <u>top of the image</u>.

6) Repeat the process for a point on the <u>bottom</u> of the object. When the bottom of the object is on the <u>axis</u>, the bottom of the image is <u>also</u> on the axis.

*Distance from the mirror affects the **image***

1) With an object <u>at C</u> (centre of curvature), you get a <u>real</u>, <u>upside down</u> image the <u>same size</u> as the object, in the <u>same place</u>.

2) <u>Between C and F</u>, you get a <u>real</u>, <u>upside down</u> image, bigger than the object and <u>behind</u> it.

3) An object <u>in front of F</u> makes a <u>virtual</u> image the <u>right way up</u>, bigger than F, <u>behind the mirror</u>.

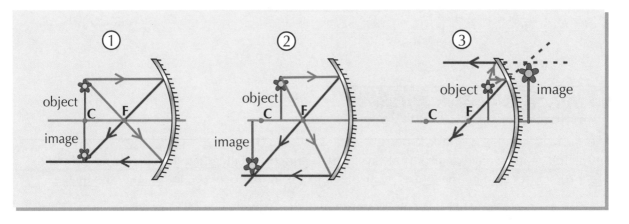

Mirrors

Draw a *ray diagram* for an *image* in a *convex mirror*

1) An incident ray <u>parallel</u> to the <u>axis</u> will reflect so that the reflected ray seems to come from the <u>focal point</u>.

2) An incident ray that can be extended to pass through the <u>focal point</u> will be <u>parallel</u> to the <u>axis</u> when it's reflected.

Always extend the lines far enough behind the mirror to be sure they pass through the focal point if they need to.

1) Pick a point on the <u>top</u> of the object. Draw a ray going from the object to the mirror <u>parallel</u> to the axis of the mirror. Make it a <u>bold line</u> when it's in front of the mirror, and a <u>dotted line</u> behind.

2) Draw another line going from the top of the <u>object to the mirror</u>, passing through the <u>focal point</u> on the other side. Make it <u>dotted</u> when it's <u>behind</u> the mirror.

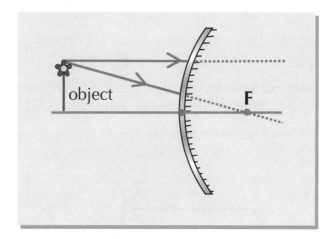

3) The incident ray that's <u>parallel</u> to the axis is <u>reflected</u> as if it starts at the <u>focal point</u>. Make sure the reflected ray <u>meets the incident ray</u> at the mirror surface.

4) The incident ray that passes through the <u>focal point</u> is reflected <u>parallel</u> to the axis. Make sure the reflected ray <u>meets the incident ray</u> at the mirror surface.

5) Mark where these two reflected rays meet behind the mirror. That's the <u>top of the image</u>.

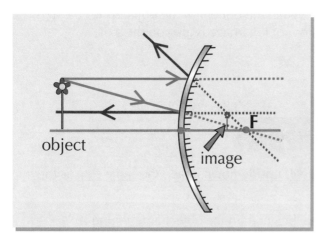

6) Repeat the process for a point on the <u>bottom of the object</u>.

The *image* is always *smaller*

1) The image in a convex mirror is always <u>virtual</u>, <u>upright</u>, <u>smaller than the object</u> and <u>behind the mirror</u>, <u>closer than F</u>. The <u>further away</u> the <u>object</u> is from the mirror, the <u>smaller</u> the <u>image</u>.

2) You can <u>see a wide area</u> in a convex mirror, which is why they put them on dodgy road corners.

Warm-Up and Exam Questions

Time for some questions — if you think this section's firmly in your head, then have a go at these...

Warm-Up Questions

1) What is the difference between a real image and a virtual image?
2) State the relationship between the angles of incidence and reflection for a reflected ray.
3) Draw a ray diagram for the reflection of an object in a plane mirror. Label the rays.
4) Draw a diagram of a convex mirror, showing its focal point, centre of curvature, axis and vertex.
5) What three things do you need to say to fully describe an image?

Exam Questions

1   Jennie has bought a new car.

(a) The car has wing mirrors on both sides so that Jennie can see what's behind her while she is driving. The wing mirrors have the following notice on them:
"Warning! Objects seen in this mirror are larger than they appear."

What type of mirror has been used for the wing mirrors?

(1 mark)

(b) In each of the car's headlights, there is a concave reflector behind the bulb.

  (i) If the focal point of the reflector is 3 cm in front of the vertex, what is the distance of the centre of curvature from the vertex?

(1 mark)

  (ii) The bulb is positioned at the focal point of the reflector. Complete the ray diagram below to show how this gives a beam of parallel light rays.

(3 marks)

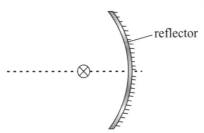

(c) Use your knowledge of reflection to explain why, before she gets in the car, Jennie can see her reflection clearly in the car's paintwork but not at all in its dull plastic bumper.

(2 marks)

2   Gregor has dropped his mobile phone in the garden pond. When he reaches in to get it, he is surprised to find that the pond is deeper than it looks. Explain why this is, and name the phenomenon that causes this effect.

(3 marks)

3   A pencil is placed in front of a convex mirror. Complete the ray diagram to the right to show how the image of the pencil is formed. The black dot on the diagram is the focal point.

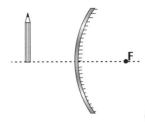

(4 marks)

Refractive Index and Snell's Law

So you're happy with the last few pages. And you're sure about that. Good. Gets a bit hairy here...

Every transparent material has a refractive index

1) The <u>absolute refractive index</u> of a material is defined as:

$$\text{refractive index, } n = \frac{\text{speed of light in a vacuum, } c}{\text{speed of light in that material, } v}$$

$$n = \frac{c}{v}$$

(Remember — the speed of light in a vacuum, $c = 3 \times 10^8$ m/s)

2) Light <u>slows down a lot</u> in <u>glass</u>, so the <u>refractive index</u> of glass is <u>high</u> (around 1.5). The refractive index of <u>water</u> is a bit <u>lower</u> (around 1.33) — so light doesn't slow down as much in water as in glass.

3) The <u>speed of light in air</u> is about the <u>same</u> as in a <u>vacuum</u>, so the <u>refractive index</u> of <u>air</u> is 1 (to 2 d.p.).

4) According to Snell's law, the <u>angle of incidence</u>, <u>angle of refraction</u> and <u>refractive index</u> are all <u>linked</u>.

Snell's Law says...

<u>When an incident ray passes into a material</u>:

$$n = \frac{\sin i}{\sin r}$$

So if you know <u>any two</u> of <u>n</u>, <u>i</u> or <u>r</u>, you can work out the <u>missing one</u>.

(Thankfully you don't have to know <u>why</u> Snell's law works. Just that it does.)

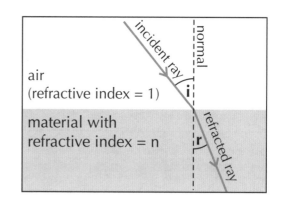

air
(refractive index = 1)

material with
refractive index = n

Example

A beam of light travels from air into water.
The angle of incidence is 23°.

Calculate the angle of refraction to the nearest degree.

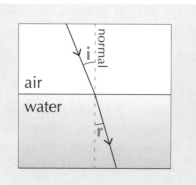

air

water

<u>Answer</u>: Using Snell's law, $\sin r = \dfrac{\sin i}{n} = \dfrac{\sin 23°}{1.33} = 0.29$

$r = \sin^{-1}(0.29) = \underline{17°}$

Refractive Index and Snell's Law

Refractive index explains *dispersion...*

1) The <u>refractive index</u> of a medium is the <u>ratio</u> of the speed of light in a vacuum to the speed of light in that medium.

2) So any material has a <u>different refractive index</u> for each <u>different speed of light</u>.

3) <u>Red</u> light <u>slows down</u> the <u>least</u> when it travels from air into glass, so it is refracted the least and has the <u>lowest refractive index</u>. <u>Violet light</u> has the <u>highest refractive index</u>.

> Here's how the refractive index changes with the wavelength of light for a certain type of glass:
>
> | colour of light | red | yellow | blue | violet |
> |---|---|---|---|---|
> | wavelength (in nm) | 656 | 589 | 486 | 434 |
> | refractive index of glass | 1.514 | 1.517 | 1.523 | 1.528 |

4) This of course produces the famous <u>dispersion</u> effect — see page 200.
(It's the same with rainbows — they're due to the different refractive indices of water for different colours.)

You can use *Snell's law* to find *critical angles*

Yes yes — you've done this on page 54, but now you get all the details and a pretty equation too.

1) When light leaves a material with a <u>higher refractive index</u> and enters a material with a <u>lower refractive index</u>, it <u>speeds up</u> and so bends <u>away from the normal</u> — e.g. when travelling from <u>glass into air</u>.

2) If you keep <u>increasing</u> the <u>angle of incidence</u>, the <u>angle of refraction</u> gets closer and closer to <u>90°</u>. Eventually i reaches a <u>critical angle</u> C for which <u>r = 90°</u>. The light is refracted right along the <u>boundary</u>.

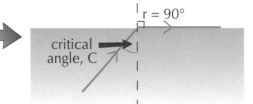

3) Above this critical angle, you get <u>total internal reflection</u> — no light leaves the medium.

4) You can find the <u>critical angle</u>, C, using this equation:

$$\sin C = \frac{n_r}{n_i}$$

n_r is the <u>refractive index</u> of the stuff the light's travelling <u>TOWARDS</u>.

n_i is the <u>refractive index</u> of the material the light starts <u>FROM</u>.

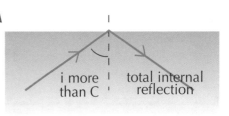

5) The <u>higher the refractive index</u>, the <u>lower the critical angle</u>. For water, C is 49°.

Refractive Index and Snell's Law

Here are a couple of <u>examples</u> to show you the type of thing they might ask, and the type of thing you should say in your answer...

Example 1

Jacob does an experiment to find the refractive index of his strawberry flavoured jelly. He finds that the critical angle for a light beam travelling from his jelly into air is 42°.

Calculate the refractive index of Jacob's jelly.

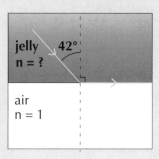

<u>Answer</u>: $\sin C = \dfrac{n_{air}}{n_{jelly}}$, so $n_{jelly} = \dfrac{n_{air}}{\sin C}$

$$\sin C = \sin 42° = 0.72$$

$$n_{air} = 1 \text{ (see page 196)}$$

so $n_{jelly} = \dfrac{1}{0.72} = \underline{1.39}$ (2 d.p.)

Example 2

Calculate the critical angle for a light beam travelling from glass (with a refractive index of 1.52) into air. Give your answer to the nearest whole degree.

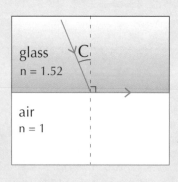

<u>Answer</u>: $\sin C = \dfrac{n_{air}}{n_{glass}} = \dfrac{1}{1.52} = 0.66$

$$C = \sin^{-1}(0.66) = \underline{41°}$$

It's always a good idea to do a sketch, so you can see clearly which numbers need to go where in the calculation.

Well — at least you've only three formulas to learn...

In <u>optical fibres</u> made of <u>glass</u> (see p.54), the <u>fibre</u> is so <u>narrow</u> that <u>light</u> signals passing through it <u>always</u> hit the boundary at angles <u>higher than C</u> — so the light is <u>always totally internally reflected</u>. It only <u>stops working</u> if the fibre is bent <u>too sharply</u>.

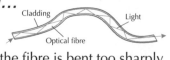

Warm-Up and Exam Questions

Another page of questions for your enjoyment. Make sure you can do them all properly...

Warm-Up Questions

1) State Snell's Law.
2) State the equation for working out the refractive index of a material using speeds.
3) Draw a diagram of a ray passing from one material into another at an angle of incidence, i.
   Label the important angles and rays.
4) Which colour of visible light is refracted the most when it travels from air into glass?
5) State the equation for calculating the critical angle for a boundary between two materials.

Exam Questions

1   Joe is doing an experiment to measure the refractive index of various materials.

   (a) First he uses an acrylic block. He shines a light ray into the block and records the
        angles of incidence and refraction. These are his results:

Angle of incidence	54°
Angle of refraction	33°

   Calculate the refractive index of acrylic from his results.

   (1 mark)

   (b) He repeats the experiment with a block of glass. He calculates the refractive index to
        be 1.56. The angle of refraction was 25°. Work out the angle of incidence.

   (2 marks)

   (c) He decides to check the accuracy of his experiment using values from a data book.
        He finds the following information:

Speed of light in glass	1.89×10^8 m/s
Speed of light in a vacuum	3.00×10^8 m/s

   Calculate the refractive index of glass from this information.

   (1 mark)

2   Marina is diving for a coin at the bottom of a swimming pool. She realises that if she
     looks up towards the surface of the water she can see the floor of the swimming pool —
     some of the light from the pool must be reflecting off the water-air boundary.

   (a) Calculate the critical angle for light coming from the pool into the air.
        The refractive index of water relative to air is approximately 1.33.
        Give your answer correct to the nearest degree.

   (3 marks)

   (b) What is the full name of the effect that occurs when a ray of light hits the
        water-air boundary at an angle greater than the critical angle?

   (1 mark)

   (c) Give a practical application of this effect.

   (1 mark)

Lenses

Lenses are usually made of glass or plastic. All lenses change the direction of light rays by refraction.

Light is refracted when it enters and leaves glass prisms

You can't fail to remember the 'ray of light through a rectangular glass block' trick:

1) The ray bends towards the normal as it enters the denser medium, and away from the normal as it emerges into the less dense medium.

   Try to visualise the shape of the wiggle in the diagram — that can be easier than remembering the rule in words.

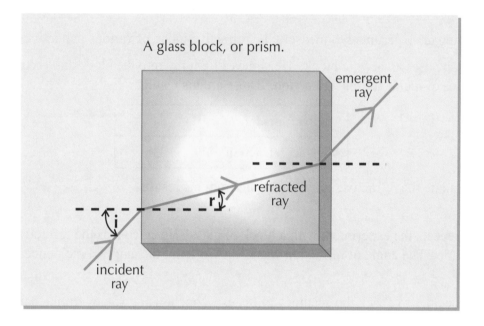

A glass block, or prism.

2) Note that different wavelengths of light refract by different amounts. So white light disperses into different colours as it enters a prism. A rectangular prism has parallel boundaries, so the rays bend one way as they enter, and then bend back again by the same amount as they leave — so white light emerges.

   But with a triangular prism, the boundaries aren't parallel, which means the different wavelengths don't recombine and you get a nice rainbow effect.

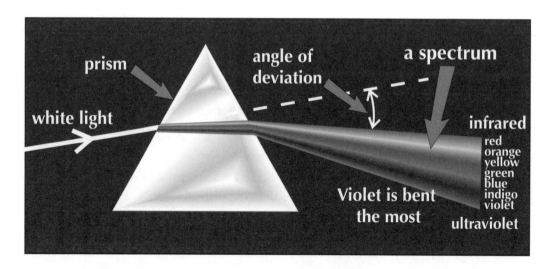

Lenses

Different lenses produce different kinds of image

There are two main types of lens — converging and diverging. They have different shapes and have opposite effects on light rays.

1) A converging lens is convex — it bulges outwards. It causes parallel rays of light to converge (move together) to a focus.

2) A diverging lens is concave — it caves inwards. It causes parallel rays of light to diverge (spread out).

3) The axis of a lens is a line passing through the middle of the lens.

4) The focal point of a converging lens is where rays hitting the lens parallel to the axis all meet.

5) The focal point of a diverging lens is the point where rays hitting the lens parallel to the axis appear to come from — you can trace them back until they all seem to meet up at a point behind the lens.

6) Each lens has a focal point in front of the lens, and one behind.

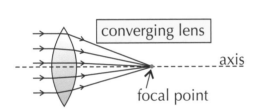

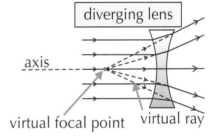

There are three rules for refraction in a converging lens...

1) An incident ray parallel to the axis refracts through the lens and passes through the focal point on the other side.

2) An incident ray passing through the focal point refracts through the lens and travels parallel to the axis.

3) An incident ray passing through the centre of the lens carries on in the same direction.

See next page for more on this.

...and three rules for refraction in a diverging lens

1) An incident ray parallel to the axis refracts through the lens, and travels in line with the focal point (so it appears to have come from the focal point).

2) An incident ray passing towards the focal point refracts through the lens and travels parallel to the axis.

3) An incident ray passing through the centre of the lens carries on in the same direction.

See page 203 for more on this.

The good thing about these rules is that they allow you to draw ray diagrams without bending the rays as they go into the lens and as they leave the lens. You can draw the diagrams as if each ray only changes direction once, in the middle of the lens.

Converging Lenses

You may have to draw a ray diagram of <u>refraction through a lens</u>. Follow the instructions very carefully...

Draw a **ray diagram** for an **image** through a **converging lens**

1) Pick a point on the <u>top</u> of the object. Draw a ray going from this point to the lens <u>parallel</u> to the axis of the lens.

2) Draw another ray from the top of the object going right through the <u>middle</u> of the lens.

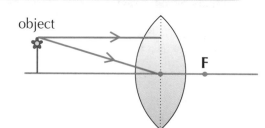

3) The incident ray that's <u>parallel</u> to the axis is <u>refracted</u> through the <u>focal point</u>. Draw a <u>refracted ray</u> passing through the <u>focal point</u>.

4) The ray passing through the <u>middle</u> of the lens doesn't bend.

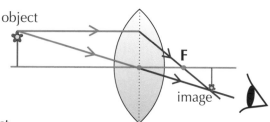

5) Mark where the rays <u>meet</u>. That's the <u>top of the image</u>.

6) Repeat the process for a point on the bottom of the object. When the bottom of the object is on the <u>axis</u>, the bottom of the image is <u>also</u> on the axis.

> If you <u>really</u> want to draw a <u>third incident ray</u> passing through the <u>focal point</u> on the way to the lens, you can (refract it so that it goes <u>parallel to the axis</u>). In the <u>exam</u>, you can get away with <u>two rays</u>, so no need to bother with three.

Distance from the lens affects the **image**

1) An object <u>at 2F</u> will produce a <u>real</u>, <u>upside down</u> image the <u>same size</u> as the object and <u>at 2F</u>.

2) <u>Between F and 2F</u> it'll make a <u>real</u>, <u>upside down</u> image <u>bigger</u> than the object and <u>beyond 2F</u>.

3) An object <u>nearer than F</u> will make a <u>virtual</u> image the <u>right way up</u>, <u>bigger</u> than the object, on the <u>same side</u> of the lens and <u>further away than 2F</u>.

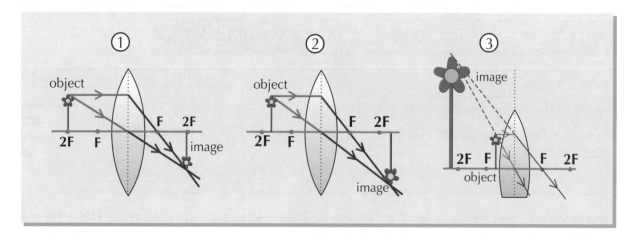

Diverging Lenses

Draw a *ray diagram* for an *image* through a *diverging lens*

1) Pick a point on the <u>top</u> of the object. Draw a ray going from the object to the lens <u>parallel</u> to the axis of the lens.

2) Draw another ray from the top of the object going right through the <u>middle</u> of the lens.

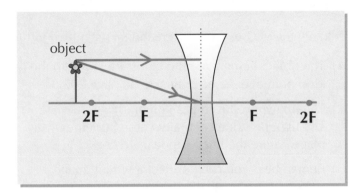

3) The incident ray that's <u>parallel</u> to the axis is <u>refracted</u> so it appears to have come from the <u>focal point</u>. Draw a <u>ray</u> from the focal point. Make it <u>dotted</u> before it reaches the lens.

4) The ray passing through the <u>middle</u> of the lens doesn't bend.

5) Mark where the refracted rays <u>meet</u>. That's the top of the image.

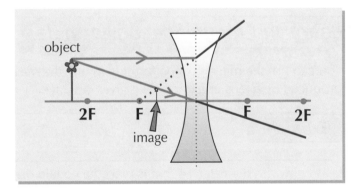

6) Repeat the process for a point on the bottom of the object. When the bottom of the object is on the <u>axis</u>, the bottom of the image is <u>also</u> on the axis.

Again, if you <u>really</u> want to draw a <u>third incident ray</u> in the direction of the <u>focal point</u> on the far side of the lens, you can. Remember to refract it so that it goes <u>parallel to the axis</u>. In the <u>exam</u>, you can get away with <u>two rays</u>. Choose whichever two are easiest to draw — don't try to draw a ray that won't actually pass through the lens.

The *image* is always *virtual*

1) A diverging lens always produces a <u>virtual image</u>.

2) The image is the <u>right way up</u>, <u>smaller</u> than the object and on the <u>same side of the lens as the object</u> — <u>no matter where the object is</u>.

Uses — Magnification and Cameras

Converging lenses are used in <u>magnifying glasses</u> and in <u>cameras</u>.

Magnifying glasses use **convex lenses**

Magnifying glasses work by creating a <u>magnified virtual image</u>.

1) The object being magnified must be closer to the lens than the <u>focal length</u> (or you get a different kind of image — see diagrams on page 202).

2) The image produced is a <u>virtual image</u>. The light rays don't <u>actually</u> come from the place where the image appears to be.

3) Remember, 'you <u>can't</u> project a virtual image onto a screen' — that's a <u>useful phrase</u> to use in the exam if they ask you about virtual images.

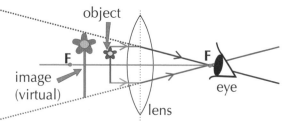

Learn the **magnification formula**

You can use the <u>magnification formula</u> to find the magnification produced by a <u>lens</u> or a <u>mirror</u> at a given distance:

$$\text{magnification} = \frac{\text{image height}}{\text{object height}}$$

Example

A coin with diameter 14 mm is placed a certain distance behind a magnifying lens. The virtual image produced has a diameter of 35 mm. What is the magnification of the lens at this distance?

<u>Answer</u>: magnification = 35 ÷ 14 = 2.5

In the exam you might have to draw a <u>ray diagram</u> to show where an image would be, and then <u>measure the image</u> so that you can work out the magnification of the lens or mirror. Another reason to draw those ray diagrams <u>carefully</u>...

Taking a **photo** forms an **image** on the **film**

When you take a photograph of a flower, light from the object (flower) travels to the camera and is refracted by the lens, forming an image on the film.

1) The image on the film is a <u>real image</u> because light rays actually meet there.

2) The image is <u>smaller</u> than the object, because the object's <u>further away</u> than the <u>focal length</u> of the lens.

3) The image is <u>inverted</u> — upside down.

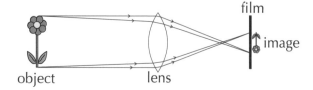

4) The <u>same</u> thing happens in our <u>eye</u> — a <u>real, inverted image</u> forms on the <u>retina</u>. Our very clever brains <u>flip</u> the image so that we see it the right way up.

Warm-Up and Exam Questions

Lots of questions on lenses here. Go on, answer them, you know you want to...

Warm-Up Questions

1) Name the two main types of lens.
2) Of the two types of lens, which bulges outwards at the centre and which curves inwards?
3) Of the two types of lens, which always creates a virtual image and which can create both real and virtual images?
4) What is meant by the focal point of a diverging lens?
5) State the magnification formula.
6) Give two uses for a convex lens.

Exam Questions

1   (a) Using this diagram of a triangular prism, explain what is meant by dispersion, and why it happens.

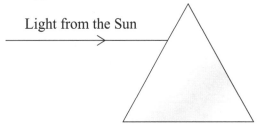

Light from the Sun

(4 marks)

     (b) Why doesn't the dispersion effect occur with a rectangular glass block?

(1 mark)

2   (a) Edward is trying to start a campfire by focussing sunlight through his spectacle lens onto the firewood. The lens is concave. Explain why he cannot focus the sunlight onto the wood using this lens.

(2 marks)

     (b) Edward finds a slug and uses a magnifying glass to look at it.

        (i) Complete the ray diagram below to show how the image of the slug is formed.

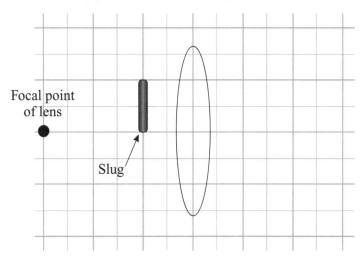

Focal point of lens

Slug

(3 marks)

       (ii) What is the magnification of the lens for the slug at this distance?

(1 mark)

Interference of Waves

Waves can <u>interfere</u> with each other, you know. If you've got two big speakers in a hall, you can get areas of loud and quiet bits, where the waves have either <u>added</u> to each other or <u>cancelled out</u>.

When **waves meet** they cause a **disturbance**

1) All waves cause some kind of <u>disturbance</u> in a medium — water waves disturb water particles, sound waves disturb air particles, electromagnetic waves disturb electric and magnetic fields.

2) When <u>two waves meet</u> at a point they both try to cause their own disturbance.

3) Waves either disturb in the <u>same direction</u> (<u>constructive</u> interference), or in <u>opposite directions</u> (<u>destructive</u> interference).

4) Think of a '<u>pulse</u>' travelling down a slinky spring meeting a pulse travelling in the opposite direction. These diagrams show the <u>possible outcomes</u>:

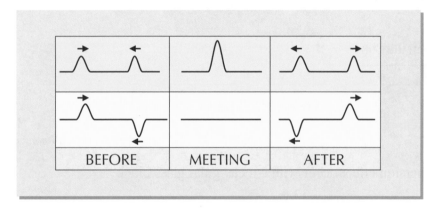

5) The <u>total amplitude</u> of the waves at a point is the <u>sum</u> of the <u>displacements</u> (you have to take direction into account) of the waves at that point.

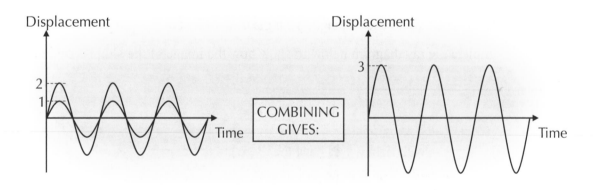

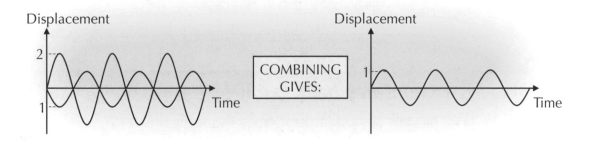

Interference of Waves

You get *patterns* of *'loud'* and *'quiet'* bits with *sound*

Two speakers both play the same note, starting at <u>exactly</u> the <u>same time</u>, and are arranged as shown:

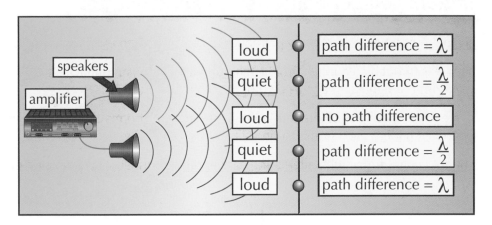

Depending on where you stand in front of them, you'll either hear a <u>loud sound</u> or <u>almost nothing</u>.
Here's why:

1) At certain points, the sound waves will be <u>in phase</u> — here you get <u>constructive interference</u>.
   The <u>amplitude</u> of the waves will be <u>doubled</u>, so you'll hear a <u>loud sound</u>.

2) These points occur where the <u>distance travelled</u> by the waves from both speakers is either the
   <u>same</u>, or different by a <u>whole number of wavelengths</u>.

3) At certain other points the sound waves will be exactly <u>out of phase</u> — here you get <u>destructive
   interference</u> and the waves will <u>cancel out</u>. This means you'll hear almost <u>no sound</u>.

4) These out of phase points occur where the difference in the <u>distance travelled</u> by the waves
   (the 'path difference') is ½ wavelength, 1½ wavelengths, 2½ wavelengths, etc.

Interference of *light* makes *'bright'* and *'dark'* bits

1) <u>Observing interference</u> effects with <u>light</u> waves is really <u>difficult</u> because their wavelengths are
   so <u>small</u>. <u>Path differences</u> with light waves have therefore got to be <u>really tiny</u>.

2) A chap called <u>Young</u> managed it by shining light through a pair of <u>narrow slits</u> that were
   just a <u>fraction of a millimetre</u> apart. This light then hit a screen in a dark room. There was an
   interference pattern of <u>light bands</u> (constructive) and <u>dark bands</u> (destructive) on the screen.

3) He was able to calculate the <u>wavelength of light</u> and help physicists unravel some of its mysteries.

Young did his experiment way back in 1801

It's weird, isn't it... I mean, constructive interference makes perfect sense — two waves, bigger sound...
it's just destructive interference that gets me. I know I've just drawn the diagrams — I know WHY it
happens... but I still find it weird. Just one of the Universe's little quirks, I suppose. (Like peanut butter.)

Diffraction Patterns and Polarisation

When light <u>diffracts</u> (spreads out through a gap) it also makes an <u>interference pattern</u> — just to confuse everyone. This is a bit tricky, but it's pretty interesting (and useful) so listen up...

When *light diffracts* you get *patterns* of *light* and *dark*

1) You get <u>interference patterns</u> when waves of <u>equal frequency</u> or <u>wavelength</u> <u>overlap</u>.

2) When a wavefront passes through a <u>gap</u>, <u>light</u> from <u>each point</u> along the gap <u>diffracts</u>. It's as if <u>every point along the wavefront is a light source in its own right</u>. Strange but true.

3) <u>Diffracted light</u> from <u>each</u> of these points interferes with light diffracted from all the <u>other points</u>. So you get an <u>interference pattern</u> even from just <u>one slit</u>.

4) The pattern has a <u>bright central fringe</u> with <u>alternating dark and bright fringes</u> on either side of it.

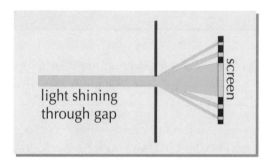

light shining through gap

screen

Transverse waves can be *plane polarised*

1) You can make a <u>transverse wave</u> by shaking a rope <u>up and down</u>, or <u>side to side</u>, or in a <u>mixture</u> of directions. Whichever <u>plane</u> you're shaking it in, it's still a transverse wave.

2) Now imagine trying to pass a rope that's waving about in <u>all different directions</u> through the slats of a wooden fence.

3) The only vibrations that'll get through the fence are the <u>vertical</u> ones. The fence <u>filters out</u> vibrations in all other directions. This is called <u>plane polarisation</u> of the wave.

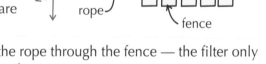

direction of waves

rope

fence

4) <u>Light</u> waves are transverse. And <u>ordinary light</u> waves are a <u>mixture of vibrations</u> in different directions.

5) Passing light through a <u>polarising filter</u> is like passing the rope through the fence — the filter only <u>transmits</u> (lets through) vibrations in one particular direction.

6) That means if you have two polarising filters at <u>right angles</u> to each other, <u>no</u> light can get through. This would be a pain if you wanted to see where you were going, but it can be quite useful:

1) <u>Polaroid sunglasses</u> act as polarising filters.

2) When light is <u>reflected</u> from a <u>horizontal surface</u> it is (partly) <u>horizontally polarised</u>.

3) So a <u>vertical polariser</u> can filter out <u>reflected glare</u> from the <u>sea</u> or the <u>snow</u> especially well.

Warm-Up and Exam Questions

There's some pretty tricky stuff in this section — interference of waves isn't the easiest
of topics — so have a good go at these questions to see if you're ready to move on...

Warm-Up Questions

1) What are the two types of interference?
2) How do you work out the total amplitude of two combined waves?
3) In an interference pattern formed by two loudspeakers, what is the difference in the path difference between two adjacent areas of loud sound (in terms of wavelength)?
4) Light waves spreading out after passing through a gap is an example of what effect?
5) What does plane polarisation mean?
6) What type of wave can be polarised?

Exam Questions

1 (a) A sound engineer is testing a pair of speaker stacks set up for a concert as shown in the diagram. She plays a test sound through the speakers. Describe the sound pattern she might hear as she walks in the direction shown by the arrow.

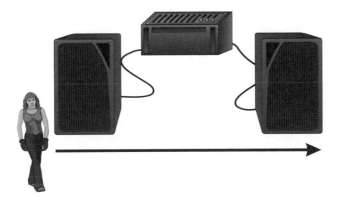

(1 mark)

   (b) What is the name of the phenomenon that causes this sound pattern?

(1 mark)

2 Why are some car windscreens made of vertically polarising glass?

(2 marks)

3 The two waves shown below are combined. On the blank grid, draw the shape of the wave created when they combine.

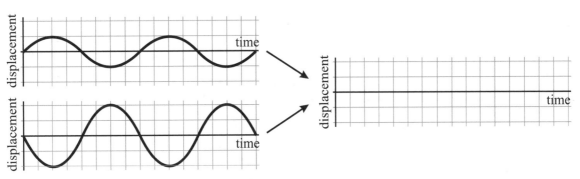

(3 marks)

Sound Waves

We hear sounds when vibrations reach our eardrums. You'll need to know how sound waves work.

Sound travels as a wave

1) Sound waves are caused by vibrating objects. These mechanical vibrations are passed through the surrounding medium as a series of compressions. They're longitudinal waves.

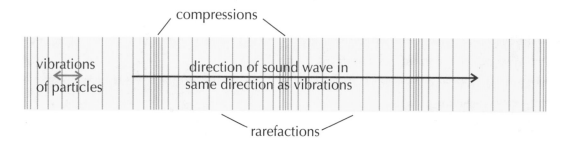

2) Sometimes the sound will eventually reach someone's eardrum, at which point the person might hear it (if it's loud enough and in the right frequency range — see next page).

3) Because sound waves are caused by vibrating particles, the denser the medium, the faster sound travels through it, generally speaking anyway. Sound generally travels faster in solids than in liquids, and faster in liquids than in gases.

> Don't get confused by CRO displays (see next pages), which show a transverse wave (like a water wave) when displaying sounds. The real sound wave is longitudinal — the display shows a transverse wave just so you can see what's going on.

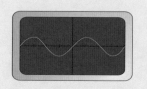

Sound waves can reflect and refract

1) Sound waves will be reflected by hard flat surfaces. Things like carpets and curtains act as absorbing surfaces which will absorb sounds rather than reflect them.

2) This is very noticeable in an empty room. A big empty room sounds completely different once you've put carpet and curtains in and a bit of furniture, because these things absorb the sound quickly and stop it echoing around the room.

3) Sound waves will also refract (change direction) as they enter different media. As they enter denser material, they speed up. (However, since sound waves are always spreading out so much, the change in direction is hard to spot under normal circumstances.)

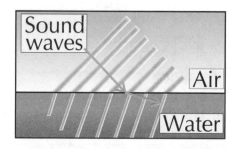

Sound Waves

We *hear* sounds in the range *20 – 20 000 Hz*

1) The frequency of a wave (in Hertz, Hz) is the number of waves in 1 second.

2) The human ear is capable of hearing sounds with frequencies between 20 Hz and 20 000 Hz.
   (Although in practice some people can't hear some of the higher frequency sounds.)

Sound does *not* travel in a *vacuum*

1) Sound waves are transmitted by vibrating particles — so they can't travel through a vacuum.
   (No particles, you see.)

2) This is nicely demonstrated by the jolly old bell jar experiment.

3) As the air is sucked out by the vacuum pump, the sound gets quieter and quieter.

4) The bell has to be mounted on something like foam to stop the sound from it travelling through the
   solid surface and making the base vibrate, because you'd hear that instead.

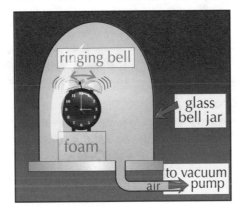

Loudness increases with *amplitude*

All sounds have pitch (see next page) and loudness. Pitch and loudness can both be measured.

1) The greater the amplitude of a wave, the more energy it carries.

2) With sound this means it'll be louder.

3) Bigger amplitude means a louder sound.

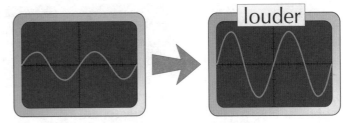

So — does a falling tree make a sound if no one hears it?...

So, we're off 'Light' and onto 'Sound' now. The thing to do here is to simply learn the facts. There's a
simple equation that says the more you learn now, the more marks you'll get in the exam. A lot of
questions just test whether you've learned the facts. Easy marks, really.

Sound Waves

*The **higher** the **frequency**, the higher the **pitch***

1) <u>High frequency</u> sound waves sound <u>high pitched</u> (like a <u>squeaking mouse</u>).
2) <u>Low frequency</u> sound waves sound <u>low pitched</u> (like a <u>mooing cow</u>).
3) <u>Frequency</u> is the number of <u>complete vibrations</u> each second.
4) Common <u>units</u> are <u>kHz</u> (1000 Hz) and <u>MHz</u> (1 000 000 Hz).
5) <u>High frequency</u> (or high pitch) also means <u>shorter wavelength</u> (see page 51).
6) These <u>CRO traces</u> are <u>very important</u>, so make sure you know them:

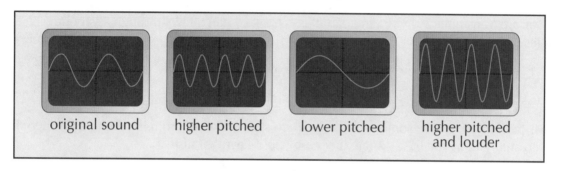

original sound   higher pitched   lower pitched   higher pitched and louder

*The **quality** of a **note** depends on the **waveform***

On a CRO trace, a clear, pure sound produces a smooth, rounded waveform called a <u>sine wave</u>.

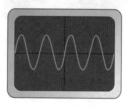

sine wave — clear, pure sound

Other kinds of sounds produce different CRO traces, for example:

1) Buzzy, brassy sounds have a <u>sawtooth waveform</u>, either with sloping 'ups' and vertical 'downs' or vertical 'ups' and sloping 'downs'.
2) A waveform of <u>rectangular peaks and troughs</u> makes a thin, <u>reedy</u> sound, a bit like an oboe.
3) A <u>square wave</u> has peaks the same length as the troughs. It makes a <u>hollow</u> sound.
4) <u>Triangle waves</u> are similar to sine waves, but they make a weaker, more <u>mellow</u> sound.

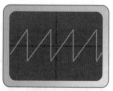 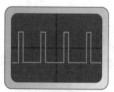 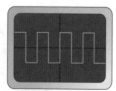 

sawtooth wave — buzzy, brassy sound   pulse wave — thin, reedy sound   square wave — hollow sound   triangle wave — weak and mellow

Waves have loudness, pitch and quality

The <u>important</u> things to remember here are what makes sounds <u>higher and lower pitched</u> and what makes sounds <u>louder and softer</u>. Once that's under your belt, have a think about the different shapes you get for different sounds. You don't need to learn all the shapes, just be aware there are differences.

Ultrasound

There's sound, and then there's <u>ultrasound</u>.

*Ultrasound is sound with a **higher frequency** than we can **hear***

1) Electrical devices can be made which produce <u>electrical oscillations</u> of <u>any frequency</u>.

2) These can easily be converted into <u>mechanical vibrations</u> to produce <u>sound</u> waves <u>beyond the range of human hearing</u> (i.e. frequencies above 20 kHz).

3) This is called <u>ultrasound</u> and it has loads of uses (see pages 215–216).

*You can use **CRO traces** to compare **amplitudes** and **frequencies***

On the screen, CRO traces of ultrasound can look just like CRO traces for <u>normal pitched</u> sounds.

For showing high frequency ultrasound, the CRO is set so that <u>each square</u> on the screen corresponds to a <u>very short time</u>, e.g. <u>1 μs</u> (0.000 001 s).

This lets you see each peak and trough:

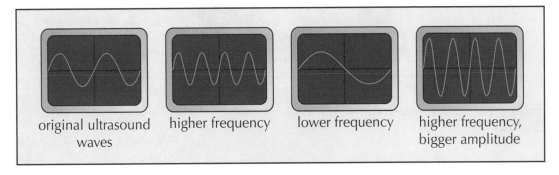

| original ultrasound waves | higher frequency | lower frequency | higher frequency, bigger amplitude |

*Ultrasound waves get **partially reflected** at a **boundary** between **media***

1) When a wave passes from one medium into another, <u>some</u> of the wave is <u>reflected</u> off the boundary between the two media, and some is transmitted (and refracted). This is <u>partial reflection</u>.

2) What this means is that you can point a pulse of ultrasound at an object, and wherever there are <u>boundaries</u> between one substance and another, some of the ultrasound gets <u>reflected back</u>.

3) The time it takes for the reflections to reach a <u>detector</u> can be used to measure <u>how far away</u> the boundary is.

4) This is how <u>ultrasound imaging</u> works — see next pages.

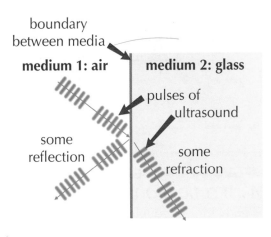

boundary between media

medium 1: air   **medium 2: glass**

pulses of ultrasound

some reflection

some refraction

Ultrasound

You can use *oscilloscope traces* to *find boundaries*

1) The CRO trace on the right shows an ultrasound pulse reflecting off <u>two separate boundaries</u>.

2) Given the 'seconds per division' setting of the CRO, you can work out the <u>time</u> between the pulses by measuring on the screen.

3) Given the <u>speed of sound</u> in the medium, you can work out the <u>distance</u> between the boundaries, using <u>d = v × t</u>.

4) They <u>might</u> give you the <u>frequency</u> and <u>wavelength</u> of the ultrasound and leave <u>you</u> to work out the speed using <u>v = frequency × wavelength</u> (see page 51).

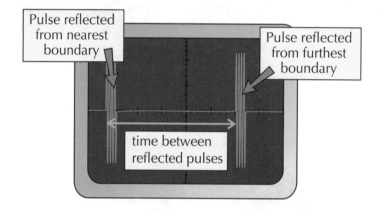

Pulse reflected from nearest boundary

Pulse reflected from furthest boundary

time between reflected pulses

Example

A pulse of ultrasound is beamed into a person's abdomen. The first boundary it reflects off is between fat and muscle. The second boundary is between muscle and a body cavity. A CRO trace shows that the time between the reflected pulses is 10 μs. The frequency of ultrasound used is 30 kHz, and the wavelength is 5 cm. Calculate the distance between the fat/muscle boundary and the muscle/cavity boundary, to give you the thickness of the muscle layer.

<u>Answer</u>: First work out the <u>speed</u> of the ultrasound using $v = f\lambda$. (Convert to Hz and m first.) $v = 30\ 000$ Hz × 0.05 m. $v = 1500$ m/s.

Next you find the distance using $d = v \times t$. BUT, the reflected pulses have travelled <u>there and back</u>, so the distance you calculate will be <u>twice the distance between boundaries</u> (think about it).

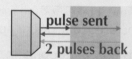

pulse sent

2 pulses back

So: $d = v \times t = 1500 \times (1 \times 10^{-5}) = 0.015$ m. So the distance between boundaries (i.e. thickness of muscle layer) = 0.015 ÷ 2 = 0.0075 m = <u>7.5 mm</u>.

With reflections you always get a factor of 2
Slightly trickier couple of pages there. Get the facts straight first — learn what <u>ultrasound</u> is, and what <u>partial reflection</u> is. Then cover the answer and make sure you can do the example question.

Ultrasound

Ultrasound has loads of exciting uses, including <u>cleaning</u>, <u>quality control</u>, <u>prenatal scans</u> and, erm, <u>bats</u>.

*Ultrasound vibrations are used in industrial **cleaning***

1) Ultrasound can be used to <u>clean delicate mechanisms</u> without them having to be dismantled.

2) Ultrasound waves can be directed onto <u>very precise areas</u>, and they're <u>extremely effective</u> at removing dirt and other deposits which form on delicate equipment.

3) The <u>high frequency vibrations</u> of ultrasound make the <u>components</u> of a piece of equipment vibrate at a high frequency. The <u>dirt</u> on the equipment vibrates too.

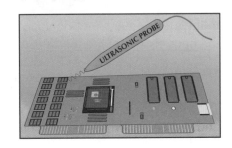

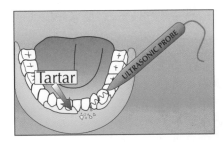

4) This vibration <u>breaks up dirt and crud</u> into very small particles, which simply fall off the equipment.

5) Alternatives to ultrasound would either <u>damage</u> the equipment (potentially), or require it to be <u>dismantled</u> before cleaning.

6) The same technique is also used by <u>dentists</u> sometimes, to clean <u>teeth</u>.

*Ultrasound is used in industrial **quality control***

1) <u>Ultrasound waves</u> can pass through something like a <u>metal casting</u>, and whenever they reach a <u>boundary</u> between <u>two different media</u> (such as metal and air), some of the wave will be <u>reflected back</u> and <u>detected</u>.

2) The exact <u>timing and distribution</u> of these <u>echoes</u> provides <u>detailed information</u> about the <u>internal structure</u> of the metal casting (or whatever).

3) The echoes are usually <u>processed by computer</u> to produce a <u>visual display</u> of what the object must be like <u>inside</u>. If there are cracks where there shouldn't be, <u>they'll show up</u>.

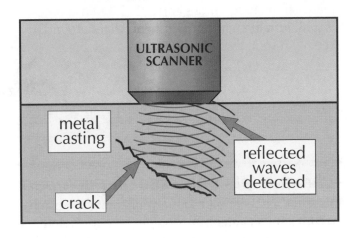

Ultrasound

*Ultrasound imaging is used for **prenatal scanning** of a foetus*

1) This follows the <u>same principle</u> as the industrial quality control. As the ultrasound hits boundaries between <u>different media</u>, some of the wave is <u>reflected</u> back.

2) In the uterus, there are boundaries between the <u>amniotic fluid</u> that the foetus floats in, and the <u>body tissues of the foetus</u> itself.

3) The reflected waves are <u>processed by computer</u> to produce a <u>video image</u> of the foetus.

4) The video image can be used to check whether the foetus is <u>developing correctly</u> — and sometimes what <u>sex</u> the foetus is too.

5) No-one knows for sure whether ultrasound is absolutely safe in all cases, but <u>X-rays</u> would definitely be dangerous to the foetus. (See page 57.)

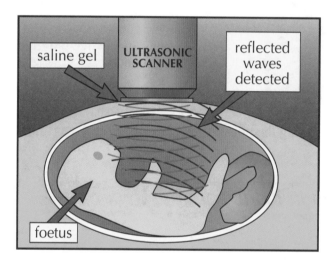

***Bats** use ultrasound to **sense** their surroundings*

<u>Bats</u> use a similar technique. They send out <u>ultrasound squeaks</u> and pick up the <u>reflections</u> with their enormous ears. Their brains <u>process</u> the reflected signals, and turn them into a <u>picture</u> of what's around. So basically, bats '<u>see</u>' with sound waves.

This ultrasound technique lets bats 'see' in pitch dark conditions

Bats are amazing, really. They can 'see' with ultrasound well enough to <u>catch a moth</u> in <u>mid-flight</u> in <u>complete darkness</u>. It's a nice trick if you can do it. <u>Another nice trick</u>, and a much easier one, is to <u>learn everything on this page</u>. Cover up the page and scribble down a mini-essay on ultrasound scans.

Warm-Up and Exam Questions

There's been lots of stuff on CRO traces and ultrasound in the last few pages — it's not too bad really, but making it stick in your head can be a bit painful... So have a pop at these questions and see how you do.

Warm-Up Questions

1) What type of wave is a sound wave?
2) Approximately what range of frequencies can humans hear?
3) Would sound travel fastest in air, a copper bar or the sea?
4) How do the frequency and amplitude of a sound wave relate to the volume and pitch of a note?
5) Give the equation relating time, distance and velocity.

Exam Questions

1   A fishing trawler is searching for shoals of fish using ultrasound sonar. The ship sends pulses of ultrasound vertically downwards and 'listens' for the echo.

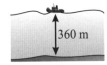

(a)   The ship is 360 m above the sea bed. A sound pulse reflected from the bottom is detected 0.48 s after it was sent from the ship. Calculate the speed of sound in sea water, showing your working.

(2 marks)

(b)   Another sound pulse is reflected off a shoal of fish. The echo is detected 180 ms after sending the pulse from the ship. How deep is the shoal of fish?

(2 marks)

(c)   Which frequency would the ship use to do this — 50 Hz, 5 kHz or 50 kHz?

(1 mark)

(d)   Give another use of ultrasound.

(1 mark)

2   Vicky is using a CRO to investigate the waveforms made by different musical instruments. She connects a microphone and amplifier to a CRO and plays the instruments in front of the microphone.

(a)   First she investigates a tuning fork. The trace on the left shows her results. Sketch the trace you would expect to see from a higher pitched tuning fork.

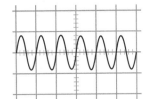   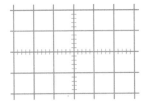

(1 mark)

(b)   Vicky now plays a note on a trumpet. The trace on the left shows her results. Sketch the trace you would expect to see if she played the same note on the trumpet, but louder.

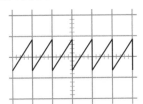   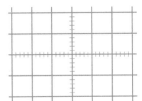

(2 marks)

Revision Summary for Section Eight

Phew, hurrah, yay — you made it to the end of this section. There are a lot of tricky ideas to remember — especially all that stuff on drawing ray diagrams for different mirrors and lenses. So get stuck into these revision questions to find out how much you've learnt...

1) A ray of light hits the surface of a mirror at an incident angle of 10° to the normal. What is the angle of reflection for this ray of light?

2)* Complete this diagram, showing
   a) the reflected rays,
   b) the position of the virtual image.

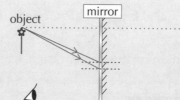

3) Sketch how the following rays of light are reflected:
   a) an incident ray parallel to the axis of a convex mirror.
   b) an incident ray passing through the focal point of a concave mirror.

4) An object is put in front of a concave mirror, between its centre of curvature and focal point. What type of image will be formed? Where will this image be?

5) Draw a diagram to show the path of a ray of light that travels from air and enters a block of glass at an angle to the normal.

6) Write down the formula for refractive index in terms of speeds of light, in words.

7) For which colour of light does glass have the highest refractive index — violet, green or red?

8)* A beam of light enters a material from the air with i = 30°. It refracts so that r = 20°. What is the refractive index of the material?

9)* Complete this ray diagram of an image formed by a diverging lens. Make sure you show at least two incident rays, the way they are refracted, and the image formed.

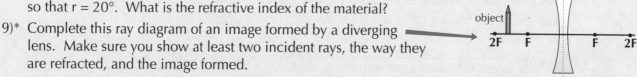

10)* Peter measures the length of a seed as 1.5 cm. When he looks at the seed through a converging lens at a certain distance, the seed appears to be 4.5 cm long. What is the magnification of the lens at this distance?

11) Describe how a sound wave travels through a medium. What type of wave is a sound wave — transverse or longitudinal?

12) Describe what happens when two sound waves constructively interfere with each other.

13) Explain why you can see an interference pattern when light is diffracted through a gap.

14) This oscilloscope trace represents a sound wave.
   a) What is happening to the volume of the sound?
   b) What is happening to the pitch of the sound?

15)* An ultrasonic scanner above the ground is used to find the depth of an underground water pipe at two different locations, A and B. For each location, the time between a pulse of ultrasound being sent from the surface and its echo being received is shown in the data table. The speed of the ultrasound is 2800 m/s. What is the depth of the pipe at the deeper of the two locations?

	Time between pulse and echo
Location A	0.002 seconds
Location B	0.003 seconds

16) Describe how ultrasound works in cleaning.

* Answers on page 279.

Potential Dividers

Potential dividers consist of a pair of resistors. They divide the potential in a circuit so you can get outputs of different voltages.

The **higher the resistance**, the **greater the voltage drop**

A voltage across a pair of resistors is 'shared out' according to their relative resistances.
The rule is:

> **The larger the share of the total resistance,
> the larger the share of the total voltage.**

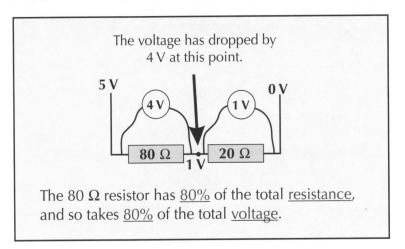

The voltage has dropped by
4 V at this point.

5 V   0 V
4 V   1 V
80 Ω   20 Ω
1 V

The 80 Ω resistor has 80% of the total resistance,
and so takes 80% of the total voltage.

The point between the two resistors is the 'output' of the potential divider.
This 'output' voltage can be varied by swapping one of the resistors for a variable resistor.

Potential dividers are quite **useful**

Potential dividers are not only spectacularly interesting — they're useful as well. They allow you to run a device that requires a certain voltage from a battery of a different voltage.
This is the formula you'll need to use:

$$V_{out} = V_{in} \times \left(\frac{R_2}{R_1 + R_2} \right)$$

Example

In the diagram, the input voltage for the potential divider is 9 V.
R_1 is 20 Ω and R_2 is 40 Ω. What is the output voltage across R_2?

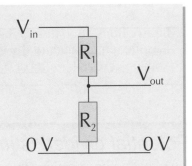

V_{in}
R_1
V_{out}
R_2
0 V   0 V

Answer: $V_{out} = 9\,V \times \left(\dfrac{40}{20 + 40} \right) = \dfrac{9\,V \times 40}{60} = 6\,V$

Potential Dividers

A *thermistor* in a *potential divider* makes a *temperature sensor*

1) Using a <u>thermistor</u> and a <u>variable resistor</u> in a potential divider, you can make a <u>temperature sensor</u> that triggers an output device at a temperature <u>you choose</u>.

2) You can make a temperature sensor that gives a <u>high voltage output</u> (a 'logical 1' — see page 226) when it's hot and a <u>low voltage output</u> (a 'logical 0') when it's cold. This is how it works...

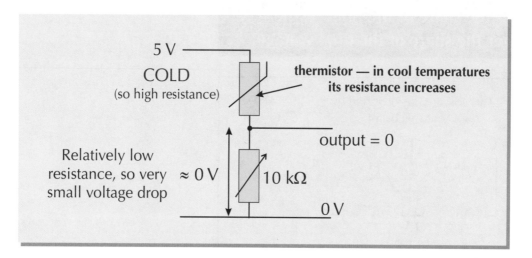

When the thermistor's <u>cold</u> its resistance is <u>very high</u>, so the voltage drop across it is <u>almost 5 V</u>, meaning the voltage of the output is <u>nearly 0 V</u> — a 'logical 0'.

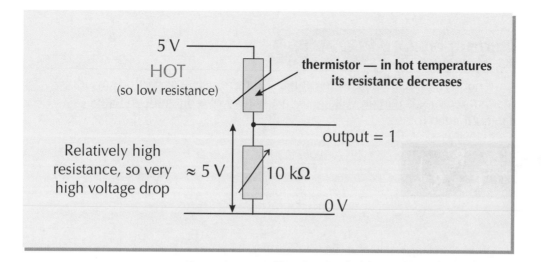

As the temperature of the thermistor <u>increases</u>, its resistance <u>falls</u> dramatically. So the voltage across it is <u>almost 0 V</u> and the voltage of the output is <u>nearly 5 V</u> — a 'logical 1'.

'Potential divider' is just the technical term for 'two resistors'

You're not going to believe this, but it's going to get <u>even more exciting</u> on the next page. I know what you're thinking — you're worried that your body won't be able to cope with the <u>adrenaline rush</u>. It's just something you have to get used to with Physics, I'm afraid.

Diodes and Rectification

Mains electricity supplies <u>alternating current</u> (AC), but many devices need <u>direct current</u> (DC). So you need a way of turning AC into DC. That's where <u>diodes</u> come in.

Diodes *only let* **current flow** *in* **one direction**

1) Diodes only let current flow freely in <u>one direction</u> — there's a very high resistance in the <u>other</u> direction.

2) This turns out to be really useful in various <u>electronic circuits</u>.

3) You can tell which direction the current flows from the <u>circuit symbol</u>.

   The <u>triangle</u> points in the direction of the current.

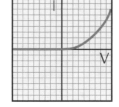

Here the current flows from <u>left to right</u>.

Diodes *are made from* **semiconductors** *such as* **silicon**

1) Diodes are often made of <u>silicon</u>, which is a <u>semiconductor</u>. This means silicon <u>can</u> conduct electricity, though not as well as a conductor.

2) Silicon diodes are made from <u>two different types</u> of silicon joined together at a '<u>p-n junction</u>'. One half of the diode is made from silicon that has an impurity added to provide <u>extra free electrons</u> — called an <u>n-type semiconductor</u> ('n' stands for the 'negative' charge of the electrons).

3) A different impurity is added to the other half of the diode so there are <u>fewer free electrons</u> than normal. There are lots of <u>empty spaces</u> left by these missing electrons which are called <u>holes</u>. This type of silicon is called a <u>p-type semiconductor</u> ('p' stands for the 'positive' charge of the holes).

4) When there's <u>no potential difference (p.d.)</u> across the diode, electrons and holes recombine across the two parts of the diode. This creates a <u>region</u> where there are <u>no holes or free electrons</u>, which acts as an <u>electrical insulator</u>.

5) When there is a p.d. across the diode the <u>direction</u> is <u>all-important</u>:

Applying a p.d. in the <u>RIGHT</u> <u>direction</u> means the <u>free holes and electrons</u> have <u>enough energy</u> to get <u>across</u> the insulating region to the other side. This means that a <u>CURRENT FLOWS</u>.

Applying a p.d. in the <u>WRONG</u> <u>direction</u> means the <u>free holes and electrons</u> are being <u>pulled the wrong way</u>, so they <u>stay</u> on the <u>same side</u> and <u>NO CURRENT FLOWS</u>.

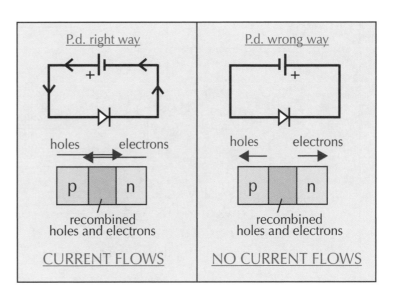

Diodes and Rectification

Diodes can be used to rectify AC current

1) A single diode only lets through current in half of the cycle. This is called <u>half-wave rectification</u>.

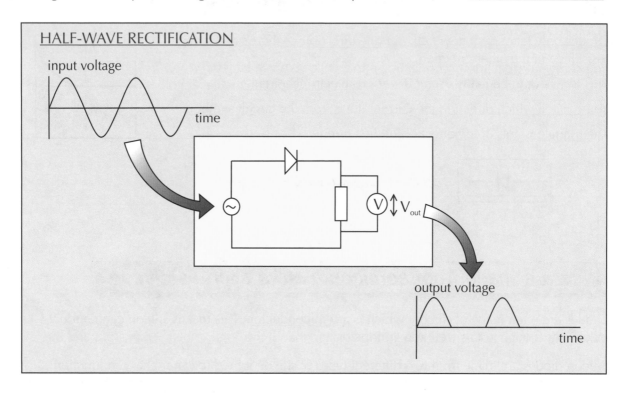

HALF-WAVE RECTIFICATION

input voltage

time

output voltage

time

2) To get <u>full-wave rectification</u>, you need a <u>bridge circuit</u> with four diodes.

In a bridge circuit, the current always flows through the component in the <u>same direction</u>, and the output voltage always has the same sign.

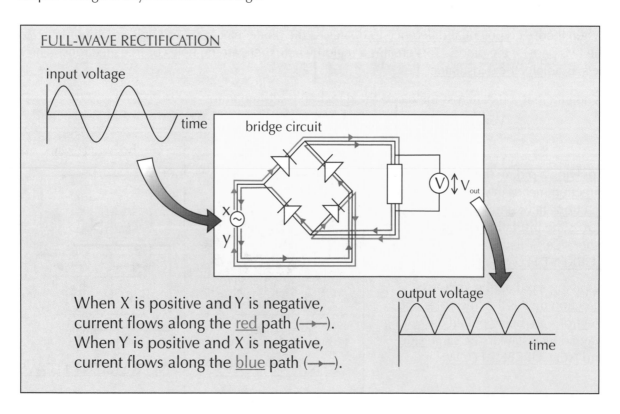

FULL-WAVE RECTIFICATION

input voltage

time

bridge circuit

output voltage

time

When X is positive and Y is negative, current flows along the <u>red</u> path (→).
When Y is positive and X is negative, current flows along the <u>blue</u> path (→).

Capacitors

AC voltage that has been rectified is not all that useful in its raw form. Chips are very sensitive to input voltage, and won't work with a voltage that looks like this: ⟋⟍⟋⟍⟋⟍. They need a <u>smoother</u> voltage like this: ∿∿. This is where <u>capacitors</u> come in handy.

Capacitors *store charge*

1) You <u>charge</u> a capacitor by connecting it to a voltage, e.g. a battery.
   A <u>current</u> flows around the circuit and <u>charge</u> gets <u>stored</u> on the capacitor.

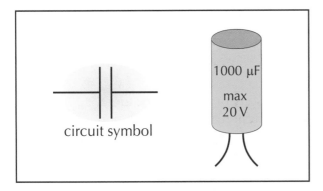

1000 µF
max
20 V

circuit symbol

2) The <u>more charge</u> that's stored on a capacitor, the <u>larger the potential difference</u> (or voltage) across it.

3) When the voltage across the capacitor is <u>equal</u> to that of the <u>battery</u>, the <u>current stops</u> and the capacitor is <u>fully charged</u>.

4) The voltage across the capacitor <u>won't rise above</u> the voltage of the battery.

5) If the battery is <u>removed</u>, the capacitor <u>discharges</u>.

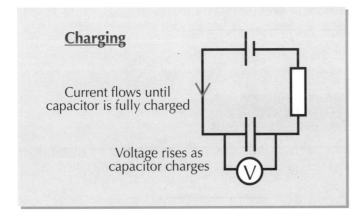

Charging

Current flows until
capacitor is fully charged

Voltage rises as
capacitor charges

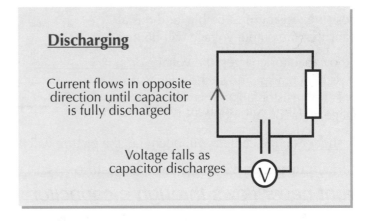

Discharging

Current flows in opposite
direction until capacitor
is fully discharged

Voltage falls as
capacitor discharges

Capacitors

Capacitors are used in 'smoothing' circuits

The output voltage from a rectified AC power supply can be 'smoothed' by adding a capacitor in parallel with the output device. A component gets current alternately from the power supply and the capacitor.

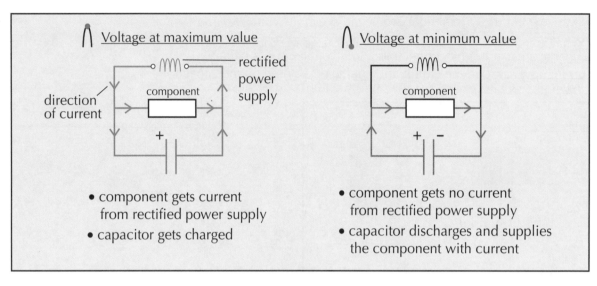

- Voltage at maximum value
  - direction of current
  - component
  - rectified power supply
  - • component gets current from rectified power supply
  - • capacitor gets charged

- Voltage at minimum value
  - component
  - • component gets no current from rectified power supply
  - • capacitor discharges and supplies the component with current

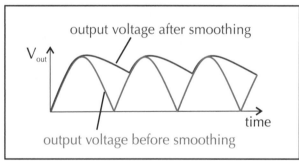

output voltage after smoothing

V_{out}

time

output voltage before smoothing

Capacitors are used to cause a time delay

Capacitors are used in timing circuits and in input sensors that need a delay. Like on a camera when you want to press the button and then run round and get in the shot before the picture's taken.

1) The switch is closed. Initially, the capacitor has no charge stored, and so the voltage drop across it is small. This means the voltage drop across the resistor must be big (and this all means that the output voltage will be low.)

2) As the capacitor charges, the voltage drop across it increases (and so the voltage drop across the resistor falls). This all means the voltage at the output increases.

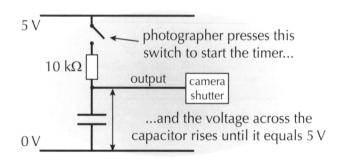

5 V

10 kΩ

output

photographer presses this switch to start the timer...

camera shutter

...and the voltage across the capacitor rises until it equals 5 V

0 V

3) The shutter on the camera will open (i.e. the picture will be taken) when the input is close to 5 V.

Current never flows through a capacitor

Capacitors just store charge, and then send current back the other way when the voltage falls.

Warm-Up and Exam Questions

OK, let's not mess about, some of this circuit stuff is pretty difficult. The best way to get it sorted is to practise, so have a go at these questions (and sneak a look at the relevant page if you need to...).

Warm-Up Questions

1) State the formula for a potential divider.
2) Draw the circuit symbol for a diode and sketch its voltage-current graph.
3) What are the two types of semiconductor used in diodes?
4) How many diodes do you need to full-wave rectify an AC current?
5) What device can you use to smooth the output of a rectifier?
6) 'A capacitor stores current.' True or false?

Exam Questions

1 The following circuit shows a potential divider.

(a) Calculate the output voltage of the potential divider, V_{out}.

(2 marks)

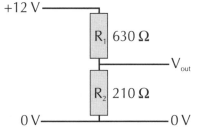

(b) How could the resistors be changed to provide a higher output voltage?

(1 mark)

(c) R_1 is replaced by a light dependent resistor (LDR). An LDR is a resistor which has a very high resistance in the dark and a low resistance in bright light.

   (i) What is the potential divider's output when it is placed in bright sunlight?

(1 mark)

   (ii) What is the potential divider's output when it is covered by thick black paper?

(1 mark)

(d) R_1 is now replaced with a diode as shown. How much current flows through R_2? Explain your answer.

(2 marks)

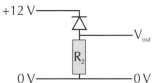

2 A capacitor is connected in a circuit with a battery, a switch and a voltmeter, as shown.

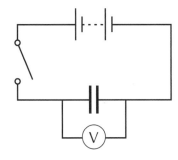

Describe what happens when the switch is closed. Refer to charge, the current that flows in the circuit and the voltage across the capacitor in your answer.

(4 marks)

Logic Gates

Learning about <u>logic gates</u> was probably how Bill Gates (no pun intended) got started. So learn all this stuff, then design a computer operating system that crashes a lot... and Bob's your uncle.

Digital systems are either on or off

1) Every connection in a digital system is in one of only <u>two states</u>. It can be either ON or OFF, either HIGH or LOW, either YES or NO, either 1 or 0... you get the picture.

2) In reality, a 1 is a <u>high voltage</u> (about 5 V) and a 0 is a <u>low voltage</u> (about 0 V). Every part of the system is in one of these two states — nothing in between.

A logic gate is a type of digital processor

<u>Logic gates</u> are small, but they're still made up of <u>lots</u> of really small components like <u>transistors</u> and <u>resistors</u>.

Each type of logic gate has its own set of <u>rules</u> for converting inputs to outputs, and these rules are best shown in <u>truth tables</u>. The important thing is to list <u>all</u> the possible <u>combinations</u> of input values.

AND and OR gates usually have two inputs

Some AND and OR gates have more than two inputs, but you don't have to worry about those.

Each input can be 0 or 1, so to allow for <u>all</u> combinations from two inputs, your table needs <u>four rows</u>.

There's a certain logic to the names...

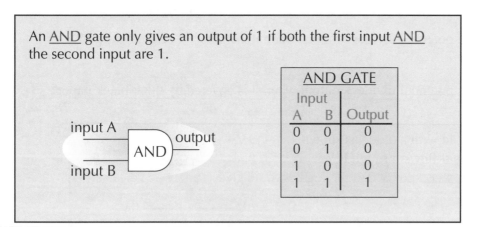

An <u>AND</u> gate only gives an output of 1 if both the first input <u>AND</u> the second input are 1.

AND GATE

Input		
A	B	Output
0	0	0
0	1	0
1	0	0
1	1	1

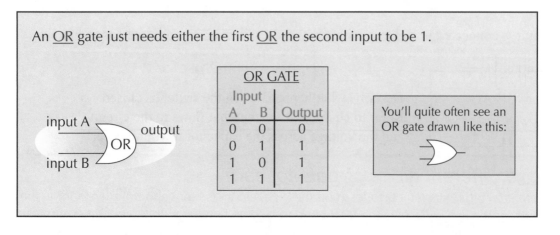

An <u>OR</u> gate just needs either the first <u>OR</u> the second input to be 1.

OR GATE

Input		
A	B	Output
0	0	0
0	1	1
1	0	1
1	1	1

You'll quite often see an OR gate drawn like this:

Logic Gates

NOT gate — sometimes called an inverter

A <u>NOT</u> gate just has <u>one</u> input — and this input can be either <u>1</u> or <u>0</u>, so the truth table has just two rows.

input — NOT — output

NOT GATE	
Input	Output
0	1
1	0

NAND and NOR gates have the opposite output of AND and OR gates

A <u>NAND gate</u> is like <u>combining</u> a <u>NOT</u> with an <u>AND</u> (hence the name):

If an AND gate would give an output of 0, a <u>NAND</u> gate would give 1, and vice versa.

input A — NAND — output
input B

NAND GATE		
Input		
A	B	Output
0	0	1
0	1	1
1	0	1
1	1	0

A <u>NOR gate</u> is like <u>combining</u> a <u>NOT</u> with an <u>OR</u> (hence the name):

If an OR gate would give an output of 0, a <u>NOR</u> gate would give 1, and vice versa.

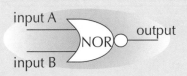

input A — NOR — output
input B

NOR GATE		
Input		
A	B	Output
0	0	1
0	1	0
1	0	0
1	1	0

There are five different types of gate to learn

Well at least there aren't that many <u>facts</u> to learn on these pages — it's more about <u>understanding</u> the inputs and outputs for the five types of gate. It's a good idea to be familiar with the circuit symbols for the gates though. And practise writing out those tables — it's the <u>best way</u> to learn them.

Using Logic Gates

You need to be able to construct a <u>truth table</u> for a <u>combination</u> of logic gates.
Approach this kind of thing in an <u>organised</u> way and <u>stick to the rules</u>, and you won't go far wrong.

'Interesting' example — a **greenhouse**

Have a look at the following example — a warning system for a <u>greenhouse</u>. Once the gardener has switched the system on, he wants to be warned if the greenhouse gets <u>too cold</u> or if <u>someone has opened the door</u>.

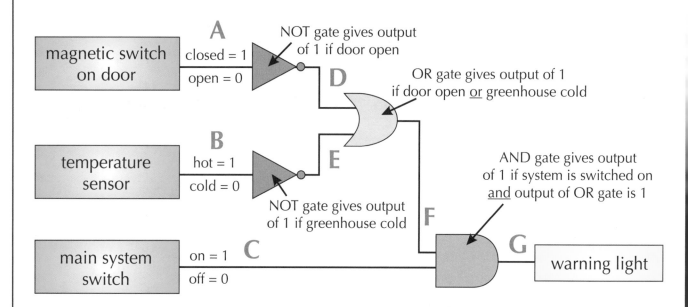

Inputs						Output
A	B	C	D	E	F	G
0	0	0	1	1	1	0
0	0	1	1	1	1	1
0	1	0	1	0	1	0
0	1	1	1	0	1	1
1	0	0	0	1	1	0
1	0	1	0	1	1	1
1	1	0	0	0	0	0
1	1	1	0	0	0	0

The <u>warning light</u> will come on if:

(i) it is <u>cold</u> in the greenhouse <u>OR</u> if the <u>door</u> is opened,

(ii) <u>AND</u> the system is switched <u>on</u>.

1) Each connection has a <u>label</u>, and <u>all</u> possible combinations of the inputs are included in the table.

2) What really matters are the <u>inputs</u> and the <u>output</u> — the rest of the truth table is just there to help.

Using Logic Gates

A *latch* works like a kind of *memory*

1) It's likely that the greenhouse will be too <u>cold</u> in the <u>middle</u> of the night but <u>warm up</u> again by <u>morning</u>. This means the warning light will have <u>gone out</u> by the time the gardener gets out of bed.

2) What the gardener needs is some way of getting the warning light to <u>stay on</u> until it is <u>seen</u> and <u>reset</u>. This is where the <u>latch</u> comes in.

3) A latch can be made by combining two <u>NOR gates</u> as shown below:

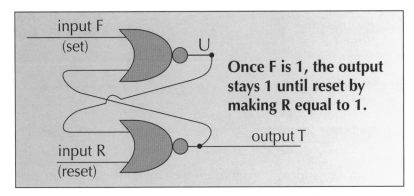

input F (set)

U

Once F is 1, the output stays 1 until reset by making R equal to 1.

output T

input R (reset)

In the system on the last page, the latch would be <u>between</u> the blue <u>OR</u> gate and the green <u>AND</u> gate.

1) *When the gardener goes to bed:*

Input F is 0 and <u>output T</u> is 0...

...meaning that the <u>top NOR gate</u> outputs 1...

...and so the <u>bottom NOR gate</u> outputs 0, which means <u>output T</u> remains 0.

2) *When the door is opened or the temperature falls:*

Input F becomes 1...

...so <u>output U</u> becomes 0...

...so the <u>bottom NOR gate</u> gives 1 (as input R is still 0), i.e. the <u>output</u> is 1.

3) *When the door is closed or the temperature rises:*

Input F becomes 0 again...

...but <u>output T</u> is 1 still...

...so <u>output U</u> stays 0 and <u>output T</u> stays 1.

4) *To reset the system:*

Briefly make <u>input R</u> equal to 1...

...and since <u>output U</u> is still 0...

...<u>output T</u> becomes 0.

LEDs and Relays in Logic Circuits

Two main points on this page: 1) An <u>LED</u> can be used to display the output of a logic gate.
2) Logic gates don't usually supply much current, so they're often connected to a more powerful circuit using a <u>relay switch</u>.

LEDs — light-emitting diodes

1) An LED is a <u>diode</u> (see page 221) which <u>gives out light</u>.

2) Like other diodes, it only lets current go through in <u>one direction</u>. When it does pass current, it gives out a pretty <u>coloured light</u>.

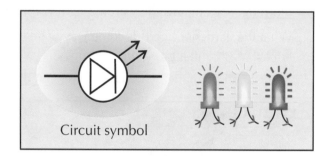

Circuit symbol

3) You can use a light-emitting diode (LED) to show the output of a <u>logic gate</u>. If the output is <u>1</u>, enough current will flow through the LED to light it up.

4) An LED is a better choice to show output than an ordinary incandescent bulb because it uses <u>less power</u> and <u>lasts longer</u>.

5) The LED is often connected in series with a <u>resistor</u> to prevent it from being damaged by too large a current flowing through it.

A **relay** is a **switch** which connects **two circuits**

1) The <u>output</u> of a logic gate usually allows only a <u>small current</u> to flow through the circuit.

2) But an <u>output device</u> like a motor requires a <u>large current</u>.

3) The solution is to have <u>two circuits</u> connected by a <u>relay</u>.

4) The relay <u>isolates</u> the <u>low voltage</u> electronic system from the <u>high voltage</u> mains often needed for the <u>output device</u>.

There are a few circuit symbols for a relay — this is the simplest one.

5) This also means that it can be made <u>safer</u> for the person <u>using</u> the device — you can make sure that <u>any parts</u> that could come into contact with a <u>person</u> are in the <u>low-current</u> circuit. For example, a <u>car's starter motor</u> needs a very <u>high current</u>, but the part <u>you control</u> (when you're turning the key) is in the <u>low-current circuit</u> — <u>safely isolated</u> by the relay.

Here's how a **relay** works...

1) When the switch in the low current circuit is <u>closed</u>, it turns on the <u>electromagnet</u> (see page 173), which <u>attracts</u> the <u>iron contact</u> on the <u>rocker</u>.

2) The rocker <u>pivots</u> and <u>closes the contacts</u> in the high current circuit — and the motor spins.

3) When the low current switch is <u>opened</u>, the electromagnet <u>stops pulling</u>, the rocker returns, and the <u>high current circuit</u> is <u>broken</u> again.

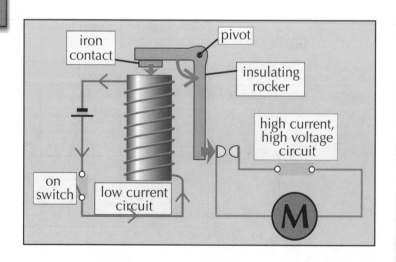

Warm-Up and Exam Questions

Here's another bunch of questions on electronics. Have a go, then read up on anything you can't do.

Warm-Up Questions

1) Write out truth tables for the following gates: AND, OR, NAND, NOR.
2) What is a latch used for? What sort of gate do latches use?
3) What does 'LED' stand for? What is the circuit symbol for an LED?
4) What is the main reason for using a relay in a logic circuit?

Exam Questions

1   (a)   Why is an LED a better choice than a filament lamp to use in an electronic circuit?

(1 mark)

(b)   How can you protect an LED in a circuit from being damaged by too high a current?

(1 mark)

2   A system that controls the temperature of a swimming pool involves a pump injecting hot water into the pool if the temperature drops too low. The pump is controlled by an electronic control circuit which operates at 5 V DC and is linked to the pump wiring by a relay. The pump runs at mains voltage.

(a)   The overall circuit diagram for the system is shown to the right. The control circuit gives either an ON output (1) or an OFF output (0).

(i)   What output will make the pump run?

(1 mark)

(ii)   Explain how the relay works.

(3 marks)

(b)   There are three inputs to the control circuit (shown right) — the main switch, a fault sensor and a temperature sensor. Complete the truth table below.

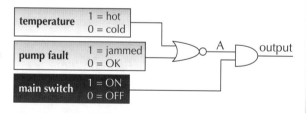

temperature	pump fault	main switch	A	output
1	1	1	0	0
1	1	0		
1	0	1		
1	0	0		
0				
0				
0				
0				

(7 marks)

Revision Summary for Section Nine

Well, that was a barrel of electrical laughs. Logic gates might not seem that logical now, but keep going through it and eventually you'll grow to love the words NAND and NOR. OK, maybe not, but at least it's a short section and there are only a few fun-filled questions to go before you've finished it.

Give 'em a bash.

1) Explain how potential dividers work.
2)* The diagram shows a potential divider with an input voltage of 9 V. R_1 is 10 Ω and R_2 is a variable resistor. What is the output voltage across R_2 when: a) $R_2 = 30$ Ω?   b) $R_2 = 2$ Ω?
3) What does a diode do?
4) What material are diodes often made of? Why?
5) Explain the two ways in which an AC current can be rectified. Include circuit diagrams and voltage/time graphs in your explanations.
6) What is a capacitor?
7) Draw a circuit diagram showing how a capacitor can be used to smooth a rectified voltage.
8) Draw a truth table for a NOT gate.
9)* The diagram below shows how logic gates can be used to monitor the temperature inside a fridge.

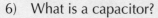

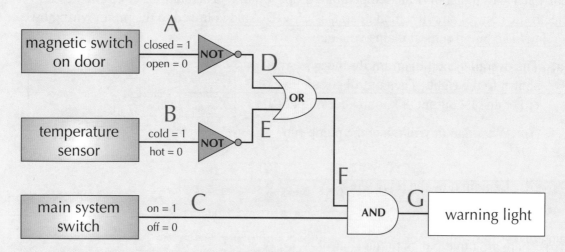

The warning light will come on if the output at G is 1. For each of the following inputs, write down whether the warning light is on or off.
a) A = 1, B = 1, C = 1
b) A = 1, B = 0, C = 1
c) A = 1, B = 1, C = 0.

10) Explain how an LED can be used to show the output of a logic gate.
11) Make a labelled sketch of a relay.

* Answers on page 280.

Kinetic Theory and Temperature in Gases

Coming up in the next couple of pages we have: <u>particles in gases</u>, <u>absolute zero</u> and the <u>Kelvin</u> scale of temperature...

Kinetic theory says gases are *randomly moving particles*

1) <u>Kinetic theory</u> says that gases consist of <u>very small particles</u>. Which they do — oxygen consists of oxygen molecules, neon consists of neon atoms, etc.

2) These particles are constantly <u>moving</u> in <u>completely random directions</u>.

3) They constantly <u>collide</u> with each other and with the walls of their container.

4) When they collide, they <u>bounce</u> off each other, or off the walls.

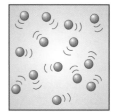

5) The particles hardly take up any space. Most of a gas is <u>empty space</u>.

Kinetic energy is proportional to *temperature*

The heading makes this sound more complicated than it actually is...

1) If you <u>increase</u> the temperature of a gas, you give its particles <u>more energy</u> — they move about more <u>quickly</u> or <u>vibrate</u> more.

2) In fact, if you <u>double</u> the temperature (measured in <u>kelvins</u>, see the next page), you <u>double</u> the average <u>kinetic energy</u> of the particles.

> The <u>temperature of a gas</u> (in <u>kelvins</u>) is proportional to the <u>average kinetic energy</u> of its <u>particles</u>.

As you <u>heat up</u> a gas, the average speed of its particles <u>increases</u>. Anything that's <u>moving</u> (for example, a bunch of particles) has kinetic energy. Kinetic energy is <u>$\frac{1}{2}mv^2$</u>, remember (see page 123).

Kinetic Theory and Temperature in Gases

Absolute zero is as cold as stuff can get — *0 Kelvin*

1) You already know that if you increase the temperature of something, you give its particles more kinetic energy. In the same way, if you cool a substance down, you're reducing the kinetic energy of the particles.

2) The coldest that anything can ever get is –273 °C — this temperature is known as absolute zero.

3) At absolute zero, atoms have as little kinetic energy as it's possible to have.

4) Absolute zero is the start of the Kelvin scale of temperature.

5) A temperature change of 1 °C is also a change of 1 kelvin. The two scales are pretty similar — the only difference is where the zero occurs.

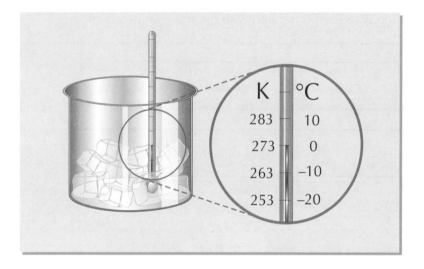

6) To convert from degrees Celsius to kelvins, just add 273. And to convert from kelvins to degrees Celsius, just subtract 273.

	Absolute zero	Freezing point of water	Boiling point of water
Celsius scale	–273 °C	0 °C	100 °C
Kelvin scale	0 K	273 K	373 K

For some reason, there's no degree symbol ° when you write a temperature in kelvins. Just write K, not °K.

7) By the way, absolute zero is actually –273.15 °C (but hardly anyone bothers about the 0.15, so you don't have to either).

And that's it — you just can't get any colder...

It's weird to think that things don't get any colder than zero kelvins (or –273 °C). I mean, what if you had a really top-of-the-range freezer that went really cold... You could set it to –273 °C, put in, say, a jelly, and then turn the freezer down a bit more. What would happen to the jelly...

Kinetic Theory and Pressure in Gases

The pressure of a gas is explained by the movement of its particles. Fascinating stuff... Well, almost.

Kinetic theory says colliding gas particles create pressure

1) As gas particles move about, they bang into each other and whatever else happens to get in the way.

2) Gas particles are very light, but they still have a (very small) mass. So when they collide with something, they exert a force on it. In a sealed container, gas particles smash against the container's walls — creating an outward pressure.

3) This pressure depends on how fast the particles are going and how often they hit the walls.

4) If you heat a gas, the particles move faster and have more kinetic energy. This extra energy means the particles hit the container walls harder and more often, creating more pressure. In fact, temperature and pressure are proportional — double the temperature and you double the pressure too.

5) And if you put the same amount of gas in a bigger container, the pressure will decrease as there'll be fewer collisions between the gas particles and the container's walls. When the volume's reduced, the particles get more squashed up and so they hit the walls more often, hence the pressure increases.

At constant volume 'P/T = constant'

Learn this equation:

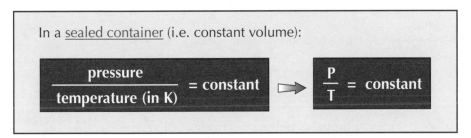

In a sealed container (i.e. constant volume):

$$\frac{\text{pressure}}{\text{temperature (in K)}} = \text{constant} \implies \frac{P}{T} = \text{constant}$$

You can also write the equation as $P_1/T_1 = P_2/T_2$ (where P_1 and T_1 are your starting conditions and P_2 and T_2 are your final conditions). Writing it like that is much more useful a lot of the time.

Example

A container has a volume of 30 litres. It is filled with gas at a pressure of 1 atm and a temperature of 290 K.

Find the new pressure if the temperature is increased to 315 K.

Answer: $P_1/T_1 = P_2/T_2$ gives:
$1 \div 290 = P_2 \div 315$
so $P_2 = 315 \div 290$

= 1.09 atm

NB: The temperatures in this formula must always be in kelvins, so if they give you the temperatures in °C, convert to kelvins FIRST. Always keep the pressure units the same as they are in the question (in this case, atm).

Kinetic Theory and Pressure in Gases

If the <u>volume</u> of the gas is <u>changing</u> (e.g. if you're compressing it or releasing it into a bigger space) then you can't use that equation on the last page. It doesn't work. Luckily there's <u>another equation</u> that you can use instead, which you can find below.

In general '(P × V)/T = constant'

If the <u>volume isn't constant</u>, this equation applies instead:

$$\frac{\text{pressure} \times \text{volume}}{\text{temperature (in K)}} = \text{constant} \qquad \Longrightarrow \qquad \frac{P \times V}{T} = \text{constant}$$

$$\text{so} \qquad \frac{P_1 V_1}{T_1} = \frac{P_2 V_2}{T_2}$$

This all applies only to so-called <u>ideal gases</u>. Ideal gases are gases that are '<u>well behaved</u>', i.e. ones that this equation works for... Scientists, eh. But don't worry, they're not going to start giving you exam questions about non-ideal gases.

Anyway, here's another <u>example</u> to show how <u>this</u> equation works:

Example

A gas at a pressure of 2.5 atmospheres is compressed from a volume of 300 cm³ down to a volume of 175 cm³. During the compression, its temperature increases from 230 K to 280 K.

Find the new pressure of the gas, in atmospheres.

<u>Answer:</u> $P_1 V_1 / T_1 = P_2 V_2 / T_2$ gives:
$(2.5 \times 300) \div 230 = (P_2 \times 175) \div 280$,
so $P_2 \times 175 = [(2.5 \times 300) \div 230] \times 280$
so $P_2 \times 175 = 913.04$
$P_2 = 913.04 \div 175$

$= \underline{5.22 \text{ atm}}$.

Less space, more collisions, more pressure

The nice thing is that you don't need to <u>fully understand</u> the physics — you just need a bit of 'common sense' about <u>formulas</u>. Understanding always helps of course, but you can still get the right answer without it. Really, you've just got to identify the values for each letter — the rest is <u>very routine</u>.

Warm-Up and Exam Questions

It's time to check those kinetic theory skills... Do these questions, and if there are any you're struggling with, go back and have a read through whichever bits your brain gave up on.

Warm-Up Questions

1) Write down the equation that links the pressure and temperature of an ideal gas at constant volume.
2) Write down the equation that links the pressure, temperature and volume of an ideal gas.
3) Why does a gas exert a pressure on the walls of its container?
4) Convert the following temperatures from degrees Celsius to kelvins:
   29 °C, −30 °C, 174 °C, 0 °C
5) Convert the following temperatures from kelvins to degrees Celsius: 200 K, 350 K, 0 K, 623 K.

Exam Questions

1   Use your ideas about kinetic theory to answer the following questions.

(a) Explain why it is dangerous to leave an aerosol in direct sunlight.

(3 marks)

(b) Sam says that a machine could be built to cool objects down to less than 0 K, provided enough power was available to run it. Is this true? Explain your answer.

(2 marks)

(c) Why do the sides of a plastic bottle collapse if you suck air out of it?

(3 marks)

2   Chris is checking his car tyre pressures. The car manufacturer says that the tyres should be checked when cool and their pressure should be set to 31 p.s.i. The temperature of the tyres is 15 °C.

(a) Why should the tyre pressures be checked when they are cool?

(1 mark)

(b) What is 15 °C in kelvins?

(1 mark)

(c) Chris sets his tyre pressures to 31 p.s.i. and goes out for a drive. When he returns he checks the pressures again and finds they are now 35 p.s.i. What is the temperature of his tyres now in degrees Celsius? (Assume the tyres have not changed volume.)

(3 marks)

3   A container holds 25 m³ of gas at a pressure of 2 atmospheres and a temperature of 20 °C. The container is compressed until the temperature reaches 60 °C. The new pressure is 2.6 atmospheres. Calculate the new volume.

(3 marks)

Particles in Atoms

You'll remember from earlier topics that the nucleus contains <u>protons and neutrons</u>. There are some <u>new types of radiation</u> on these pages — positron and neutron radiation. So don't be tempted to skim over it all.

Know the properties of **alpha**, **beta** and **gamma** radiation

To be honest, you should be on <u>first name terms</u> with <u>alpha</u>, <u>beta</u> and <u>gamma</u> already. But there's no harm in looking through the <u>basic facts again</u>, just to make sure you know what's what.

	alpha	**beta**	**gamma**
What is it?	A <u>helium nucleus</u> $^4_2$He	An <u>electron</u> $^0_{-1}$e	Electromagnetic <u>radiation</u>
Is it fast or slow, heavy or light?	<u>Slow</u> and heavy	Light and <u>fast</u>	No mass, very <u>fast</u>
Is it ionising?	<u>Strongly</u> ionising	<u>Moderately</u> ionising	<u>Weakly</u> ionising
How penetrating is it and what stops it?	Stopped by <u>paper</u>, skin, etc.	Stopped by <u>thin metal</u>	Stopped by <u>thick lead</u> or <u>very thick concrete</u>

Positron radiation is **positively charged beta** radiation $^0_{+1}$e

1) Positrons are <u>just like electrons</u> — they've got exactly the same mass — but they're <u>positively charged</u> instead of negatively charged.

2) It doesn't mean very much to talk about how <u>fast</u>, <u>ionising</u> or <u>penetrating</u> they are though, because they're <u>obliterated</u> in spectacular fashion as soon as they meet an electron. This is called <u>annihilation</u> (see page 260).

The positron is the <u>antiparticle</u> of the electron (see p.241).

Neutron radiation is... **neutrons** $^1_0$n

1) Neutrons are more penetrating than <u>alpha</u> or <u>beta</u> and sometimes even <u>more penetrating</u> than <u>gamma radiation</u>.

2) Unlike alpha, beta and gamma, neutrons aren't <u>directly ionising</u>, but they can be <u>absorbed by the nuclei</u> of atoms in the substances they pass through.

3) <u>Absorbing a neutron</u> can make a nucleus <u>radioactive</u>.

4) These radioactive nuclei then emit ionising radiation (α, β or γ), so neutrons are sometimes called '<u>indirectly ionising</u>'.

See p.184 for how neutrons act in nuclear chain reactions.

5) Neutrons are absorbed best by <u>light nuclei</u>. <u>Hydrogen</u> nuclei are the <u>lightest</u> of all, so hydrogen-rich materials such as <u>water</u>, <u>polythene</u> or <u>concrete</u> are used to make neutron radiation shielding.

6) Neutron absorption often makes nuclei emit <u>gamma radiation</u>, so some <u>nice thick lead</u> can be added to neutron radiation shielding, to shield against gamma radiation as well.

Particles in Atoms

Neutrons *are* **difficult to detect** *because they're* **neutral**

I'm sure you must be wondering why that makes a difference. Here's why:

Charged particles knock electrons off atoms they bang into, ionising them. Charged particles are often detected by looking at the trail of ions left behind. This doesn't work for neutral particles.

Charged particles also move towards charged objects. You can use an electrical field to deflect a beam of alpha particles (or anything else charged). This doesn't work for neutral particles either.

Instead, neutrons are detected by looking for nuclear decays from the nuclei they make radioactive.

Some nuclei are **more stable** than others

Radioactive materials emit alpha particles, beta particles or gamma radiation (or a combination). A nucleus will be unstable if it has:

> 1) too many neutrons
>
> 2) too few neutrons
>
> 3) too many protons and neutrons altogether, i.e. it's too heavy
>
> 4) too much energy.

Plot the number of neutrons (N) against the number of protons (Z) for stable isotopes and you get this nice curve of stability.

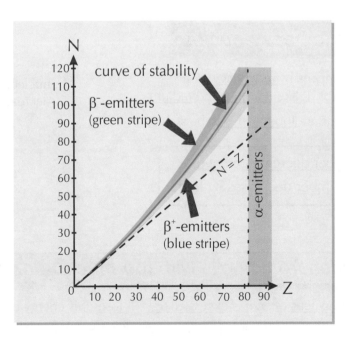

1) Any isotope which doesn't lie on the curve is unstable.
   Unstable means radioactive — particles or radiation are emitted.

2) An isotope that lies above the curve has too many neutrons to be stable.

3) An isotope that lies below the curve has too few neutrons to be stable.

Particles in Atoms

β– decay happens when there are **too many neutrons**

1) <u>Beta-minus</u> (usually just called beta) decay is the emission of an <u>electron</u> from the <u>nucleus</u>.

2) Beta decay happens in isotopes that are '<u>neutron rich</u>'
(i.e. have many more <u>neutrons</u> than <u>protons</u> in their nucleus).
When a nucleus ejects a beta particle, one of the <u>neutrons</u>
in the nucleus is <u>changed</u> into a <u>proton</u>.

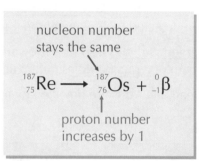

> The <u>proton number increases</u> by <u>one</u>,
> and the <u>nucleon number stays the same</u>.

β+ decay happens when there are **too few neutrons**

1) <u>Beta-plus</u> decay is the emission of a <u>positron</u> from the nucleus.

2) In <u>beta-plus emission</u>, a <u>proton</u> gets <u>changed</u> into a <u>neutron</u>.

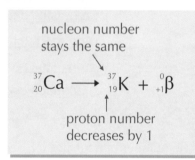

> The <u>proton number decreases</u> by <u>one</u>,
> and the <u>nucleon number stays the same</u>.

α decay happens in **heavy nuclei**

1) <u>Alpha emission</u> only happens in <u>very heavy</u> atoms
(with more than 82 protons), like <u>uranium</u> and <u>radium</u>.

2) The <u>nuclei</u> of these atoms are <u>too massive</u> to be stable.

$$^{238}_{92}U \longrightarrow ^{234}_{90}Th + ^{4}_{2}\alpha$$

nucleon number decreases by 4

proton number decreases by 2

> The <u>proton number decreases</u> by <u>two</u>,
> and the <u>nucleon number decreases</u> by <u>four</u>.

γ radiation is emitted from nuclei with **too much energy**

1) <u>After</u> α or β decay, the <u>nucleus</u> often has <u>excess energy</u>. It <u>loses</u> this energy by emitting a <u>gamma ray</u>.

2) Gamma emission always goes with beta or alpha decay. You <u>never</u> get <u>just gamma</u> rays emitted.

Just forget how strange this all is and learn the facts

OK, there is <u>a lot to chew through</u> here. Try it like this: With <u>too many</u> neutrons, you have to <u>take away</u>, so that's <u>beta minus</u>. With <u>too few</u>, you have to <u>add</u> — that's <u>beta plus</u>. For nuclear equations, balance the top line (mass) <u>AND</u> the bottom line (charge). You'll get simple ones in the exam.

Fundamental and Other Particles

You know negative <u>electrons</u> orbit a positive <u>nucleus</u> in an atom. But it's actually a bit more complex...

Electrons and positrons are fundamental particles

1) Electrons are <u>fundamental particles</u> — meaning you can't divide electrons into even <u>smaller</u> particles.

2) The electron's positive equivalent, the <u>positron</u> (see page 238) is also a <u>fundamental</u> particle.

3) It is actually possible to make new fundamental particles, odd as that may sound.

4) When <u>two protons</u> collide at very <u>high speed</u>, they release lots of <u>energy</u>. This <u>energy</u> can turn into <u>mass</u> — weird but true.

5) When <u>energy</u> is converted into <u>mass</u> you make <u>equal amounts</u> of <u>matter</u> and <u>antimatter</u>.

6) Antimatter is made of <u>antiparticles</u>. Each particle has an antiparticle which has the <u>same mass</u>, but <u>opposite charge</u>, e.g. the positron is the antiparticle of the electron.

7) Scientists can make loads of new particles in experiments like this.

Protons and neutrons are made up of smaller particles

1) <u>Protons</u> and <u>neutrons</u> are <u>NOT</u> fundamental particles. They're made up of even smaller particles called <u>quarks</u>. It takes <u>three quarks</u> to make a proton or neutron.

2) There are various kinds of quark, but protons and neutrons contain just two types — <u>up-quarks</u> and <u>down-quarks</u>.

3) A <u>proton</u> is made of <u>two up-quarks</u> and <u>one down-quark</u>.

4) A <u>neutron</u> is made of <u>two down-quarks</u> and <u>one up-quark</u>.

Don't worry about why they're called 'up' and 'down' quarks. They just are.

Quark	Relative Charge	Relative Mass
up	$\frac{2}{3}$	$\frac{1}{3}$
down	$-\frac{1}{3}$	$\frac{1}{3}$

5) The <u>relative charges</u> on up and down quarks are shown in the table. When quarks <u>combine</u> to make protons and neutrons, these charges <u>add together</u> to give the overall relative charges.

Remember: protons are 'positive' — they're more 'up' than 'down'.

| **Proton** | up-quark + up-quark + down-quark, so charge on proton $= \frac{2}{3} + \frac{2}{3} + \left(-\frac{1}{3}\right) = +1$ |

| **Neutron** | up-quark + down-quark + down-quark, so charge on neutron $= \frac{2}{3} + \left(-\frac{1}{3}\right) + \left(-\frac{1}{3}\right) = 0$ |

Quarks change — producing electrons/positrons in the process

1) Sometimes the number of <u>protons</u> and <u>neutrons</u> in a nucleus can make the nucleus <u>unstable</u>. To become more stable, a neutron is <u>converted</u> into a proton, or a proton into a neutron.

2) This happens by <u>changing</u> a down-quark into an up-quark, or vice versa.

| **Neutron** | up-quark + down-quark + down-quark ⟹ | **Proton** | up-quark + up-quark + down-quark |

3) However, the <u>overall charge</u> before and after has to be <u>equal</u>.

4) So when a <u>neutron</u> turns into a <u>proton</u>, the nucleus has to produce a <u>negatively charged</u> particle as well to keep the overall charge <u>zero</u>. The particle produced is an <u>electron</u>. This is β⁻ decay.

5) When a <u>proton</u> changes into a <u>neutron</u> (up-quark turns into <u>down-quark</u>), a <u>positive</u> charge is needed to keep the overall charge at +1. The nucleus produces and chucks out a <u>positron</u>. This is β⁺ decay.

Warm-Up and Exam Questions

There are lots of different types of decay and particles to learn here. So get answering these questions...

Warm-Up Questions

1) Name the five types of radiation and the particle or wave that is emitted in each.
2) What is a positron?
3) Why are neutrons difficult to detect?
4) Give four reasons why a nucleus may be unstable.
5) What is an antiparticle?
6) What are the relative masses and charges of the two types of quark in nuclei?
7) In terms of quarks, how is the charge made up on a neutron and on a proton?

Exam Questions

1   A team of physicists are investigating neutron radiation.

   (a) (i)   Suggest a material they could use to prevent neutron radiation escaping
              from the test equipment. Explain why your suggested material is suitable.

                                                                        (3 marks)

        (ii) Suggest another shielding material that might be required for the test
              equipment, explaining why it might be needed.

                                                                        (1 mark)

   (b) How can the physicists detect neutron radiation?

                                                                        (1 mark)

2   The diagram below shows the number of neutrons plotted against the number of protons
     for stable isotopes. The table gives information about various radioactive isotopes.

   (a) Use the graph to predict what kind of
        radioactive decay Cs-137 is likely to undergo.
        Explain your answer.
                                       (3 marks)

   (b) What kind of radioactive decay is most likely for
        U-238? Explain your answer.
                                       (2 marks)

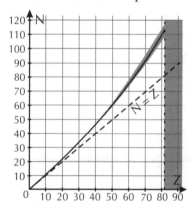

   (c) Na-22 decays by β+ decay. Complete the decay
        equation given below.

        $$^{22}_{11}\text{Na} \longrightarrow \text{............................}$$

                                       (3 marks)

   (d) Explain in terms of fundamental particles what
        happens during the β+ decay of Na-22.
                                       (3 marks)

Isotope	Nucleon no.	Proton no.
U-238	238	92
Cs-137	137	55
Mg-22	22	12
Na-22	22	11
Ne-22	22	10

Electron Beams

Electron beams are used in televisions. Electron beams come from (where else...) electron guns.

Electron guns use thermionic emission

I know this diagram looks a bit complicated, but you won't have to draw it —
you just have to understand what's going on.

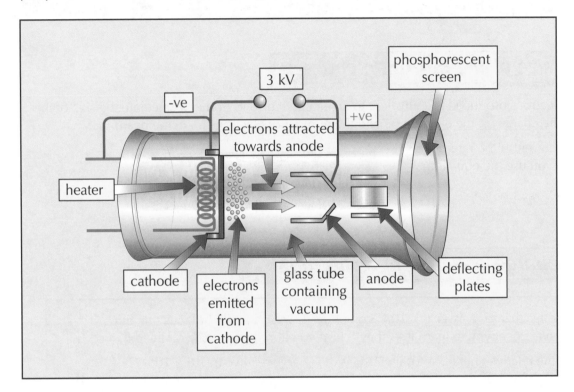

1) The heater heats the cathode, which gives more energy to its electrons. Once they have enough energy, they 'boil off', i.e. they escape.

2) This process is called thermionic emission.

3) The escaped electrons then accelerate as they're pulled towards the (positive) anode.

4) The anode has a gap in it to channel the electrons into an electron beam.

5) An electric field created between charged metal plates can be used to deflect the electron beam (see page 244).

6) The phosphorescent screen glows when the electrons hit it.

7) Electron guns in vacuum tubes are also called cathode ray tubes.

Electron Beams

You can **calculate** the **kinetic energy** of each electron...

The <u>kinetic energy</u> gained by each electron as it accelerates is given by:

kinetic energy = charge of the electron (e) × accelerating voltage (V)
(in joules, J)   (in coulombs, C)   (in volts, V)

...and the size of the **current** produced

The <u>beam</u> of electrons produced is equivalent to an electric <u>current</u>. You can calculate the size of this current using the <u>charge on the electron</u> (e) and the <u>number of electrons</u> (n) in the beam.

Remember — current is the <u>rate of flow</u> <u>of electrons</u> so you use the equation:

$$\text{current (I)} = \frac{\text{charge (n} \times \text{e)}}{\text{time (t)}}$$

Electron charge e — an **important note**...

1) The <u>electronic charge, e, is -1.6×10^{-19} coulombs (C)</u>. It's a pretty hideous number, so they <u>don't expect you to learn it</u> — they'll give you it in the exam if you need it.

2) Note that the <u>relative charge</u> (see page 241) is the charge relative to 1.6×10^{-19} C. That's arranged on purpose, of course, to make the numbers easier: electron −1, proton +1.

3) Relative charge is fine for <u>comparing charges</u> on different particles, but in <u>calculations</u> like the ones above, you need the actual value in <u>coulombs</u>.

An **electron beam** is deflected by an **electric field**

1) An electron beam is attracted by a <u>positive</u> charge and repelled by a <u>negative</u> charge.

2) Using <u>two pairs</u> of charged metal plates, the electron beam can be deflected both <u>up and down</u> (by the Y-plates) and <u>left and right</u> (by the X-plates).

3) Any stream of charged particles can be deflected like this, e.g. a jet of <u>charged ink particles</u>.

4) <u>How much</u> particles are <u>deflected</u> by depends on <u>four factors</u>:

MORE DEFLECTION	LESS DEFLECTION
<u>Bigger charge</u> on <u>plates</u> (stronger electrical field)	<u>Smaller charge</u> on <u>plates</u> (weaker electrical field)
<u>Bigger charge</u> on <u>particles</u>	<u>Smaller charge</u> on <u>particles</u>
<u>Slower</u>-moving particles	<u>Faster</u>-moving particles
<u>Lighter</u> particles	<u>Heavier</u> particles

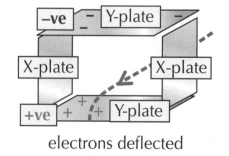

electrons deflected
towards +ve Y-plate

Electron Beams

Quite often scientists can seem to be doing things just to see if they can — but electron beams actually have several important uses that you need to learn.

Electron guns are used in *oscilloscopes*, *TVs* and to make *X-rays*

1) An oscilloscope is a device for displaying the voltage and frequency of an electrical signal. On an oscilloscope screen, the x-axis represents time and the y-axis represents voltage.

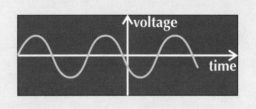

2) In television tubes, computer monitors and oscilloscopes, the electrons hit a screen covered in phosphorescent chemicals. These chemicals emit light when hit by electrons.

Plasma and LCD TVs and monitors don't have electron guns. They work differently.

3) Electron beams can also be used to produce X-rays. X-ray generators have no deflecting plates or phosphorescent screen. They have a positively-charged target instead (the anode).

When electrons hit the target, some of their kinetic energy is converted into X-rays.

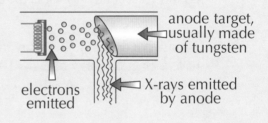

Particle accelerators help scientists find out about the Universe

And just what have particle accelerators got to do with electron guns and telly, I hear you mutter... Well, if you think about it, an electron gun also accelerates particles, just on a very small scale.

1) Scientists use huge particle accelerators to smash particles into each other at tremendous speeds, to see what happens — what kind of radiation is given off, what new particles are created, etc.

2) This gives clues about how the Universe works, so scientists can develop better explanations about the physical world.

3) Research into big scientific questions like particle physics is done internationally. Particle accelerators are so expensive that not every country can afford its own. Sharing ideas is part of science anyway.

Electron beams come from electron guns — simple

Well, I reckon this is all a bit more interesting than your average Physics topic.
I mean really — Newton's laws, circuit breakers, work done... not exactly thrilling stuff.

Warm-Up and Exam Questions

Particle accelerators and electron guns are pretty heavy science... Make sure you know your stuff.

Warm-Up Questions

1) What is the process called in which electrons escape from a heated filament?
2) What is the equation for the kinetic energy gained by an electron when it's accelerated by a voltage?
3) What is the equation for the current produced by a stream of electrons?
4) If the charge on an electron is -1.6×10^{-19} C, what is the charge in coulombs on a particle with relative charge +1?
5) In an electron gun, give four factors that result in less deflection of the electrons.
6) Give three uses of an electron gun.

Exam Questions

1   (a) Describe what happens to an electron in a cathode ray tube from before it leaves the cathode until it produces a glow on the screen. Make sure your explanation of events is in the correct order.

(4 marks)

(b) An electron gun is being used to create an electron beam. 7.5×10^{19} electrons pass a point in 5 seconds. What is the current produced by the electron beam?

(2 marks)

(c) Joanna is a dentist. The X-ray machine used at her surgery produces electrons which have a kinetic energy of 3.2×10^{-15} J. What is the accelerating voltage used by the X-ray tube? ($e = -1.6 \times 10^{-19}$ C)

(2 marks)

2   The diagram to the right shows the deflecting plates of a cathode ray oscilloscope (CRO) looking from the screen end of the tube.

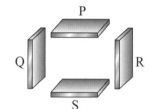

(a) (i)   Which of the traces below shows the CRO screen when a positive charge is applied to plate P?

(1 mark)

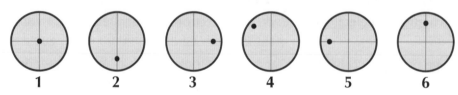

| 1 | 2 | 3 | 4 | 5 | 6 |

(ii) Which trace shows a negative charge being applied to plate R?

(1 mark)

(b) The plates are charged so that P and R are negative, and Q and S are positive. On the blank screen, draw what you would expect to see.

(1 mark)

(c) What would happen to the CRO trace if an AC voltage was applied to plates P and S?

(2 marks)

Revision Summary for Section Ten

This section covers some ordinary stuff like heating up gases, and some rather weird 'n' wonderful stuff like what happens when one kind of quark turns into another kind of quark (blimey). The real test of whether you've been paying attention is <u>in these questions</u>. You know the drill by now.

1) What's absolute zero in °C?

2) What does absolute zero mean in terms of the kinetic energy of particles?

3)* a) Convert the following temperatures into kelvins: i) –89 °C   ii) 120 °C
   b) Convert 278 K into °C.

4) How are the temperature of a gas and the kinetic energy of its particles related?

5) What two ways are there to increase the pressure of a gas in a sealed container?

6) Give the equation for the relationship between the temperature and pressure of a gas in a sealed, rigid container.

7)* A rigid container is filled with gas at a pressure of 5.0 atm. The container is cooled from 25 °C to –100 °C. What will be the new pressure inside the container?

8)* A gas is compressed from a volume of 30 dm³, a pressure of 2.0 atm and a temperature of 200 K to a volume of 10 dm³ and a pressure of 5.0 atm. What will its new temperature be?

9) a) Which is the most strongly ionising — alpha, beta or gamma radiation?
   b) Which is more ionising — alpha or positron radiation?

10) Why is neutron radiation said to be 'indirectly ionising'?

11) Particles can be detected in a device called a cloud chamber. Cloud chambers let you see the trail of ions that some particles leave behind them. Explain why neutrons can't be easily detected this way.

12) The curve of stability shows which isotopes are stable and which aren't.
    What type of radiation do nuclei just below the curve of stability emit?

13) Cobalt-60 nuclei lie above the curve of stability. What does this show about the number of neutrons in a cobalt-60 nucleus compared to number of protons? What kind of particles will cobalt-60 emit?

14) During beta-plus decay, what happens to the number of protons in the nucleus?
    What happens to the number of neutrons?

15) Uranium-238 is an alpha emitter. Write a nuclear equation to show the emission of an alpha particle by uranium-238.

16) Under what conditions is gamma radiation emitted from a nucleus?

17) Name two fundamental particles. What does it mean to say that a particle is fundamental?

18) What particles are neutrons made up of?

19) Add together two up-quarks and a down-quark. What do you get?

20) Describe what happens to the quarks in a neutron during beta-minus decay.

21) Why must an electron be produced when a neutron turns into a proton?

22) Explain how an electron gun produces a beam of electrons.

23)* Calculate the kinetic energy gained by an electron accelerated by a voltage of 230 V. ($e = -1.6 \times 10^{-19}$ C)

24)* Calculate the current produced when 2.5×10^{20} electrons flow past a point in 10 seconds.

25)* Imagine you're looking into a cathode ray tube. The top Y-plate is positively charged and the bottom Y-plate is negatively charged. The left X-plate is negatively charged and the right X-plate is positively charged. In which direction will the electron beam be deflected?

26) Why do scientists want to build massive particle accelerators?

Medical Uses of Light

Not long ago the best way for a doctor to have a look inside someone's body was to get out the scalpel, make a <u>big hole</u> and take a look. But nowadays <u>medical physics</u> has come up with better, less drastic ways to do this. Read on and learn...

Endoscopes use bundles of *optical fibres*

1) An <u>endoscope</u> is a thin tube containing <u>optical fibres</u> that lets surgeons examine <u>inside</u> the body.

2) Endoscopes consist of <u>two bundles</u> of optical fibres — one to carry <u>light</u> to the area of interest and one to carry an <u>image</u> back so that it can be viewed.

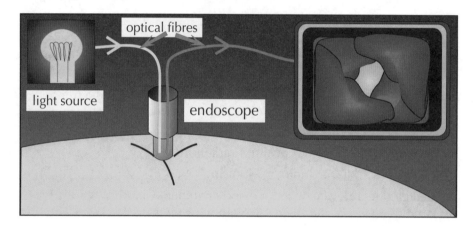

3) The <u>image</u> can be seen through an <u>eyepiece</u> or displayed as a <u>full-colour moving image</u> on a screen.

4) The <u>big advantage</u> of using endoscopes is that surgeons can now perform many <u>operations</u> by only cutting <u>tiny holes</u> in people. This is called <u>keyhole surgery</u> and it wasn't possible before optical fibres.

5) Doctors have to be careful <u>not to bend</u> the endoscope too <u>sharply</u>. If they do the light won't hit the inside surface of the optical fibre at an angle <u>greater</u> than the <u>critical angle</u>, and so <u>won't</u> be <u>reflected</u> and travel along the fibre. And then the surgeon would just be snipping away in the dark.

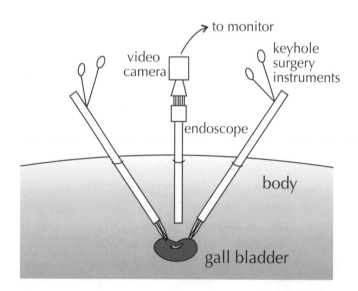

Keyhole surgery gives less risk of infection and other complications

Keyhole surgery's used for things like removing the gall bladder, investigating middle ear problems, and removing abnormal polyps in the colon. Recovery times tend to be quicker for keyhole surgery, so the patient can usually return <u>home</u> on the <u>same day</u> — cheaper for the hospital and nicer for the patient.

Medical Uses of Light

Pulse oximeters use light to check the % oxygen in the blood

Pulse oximetry measures the amount of oxygen carried by the <u>haemoglobin</u> in a patient's blood. This is useful for <u>monitoring the patient's health</u> before and after surgery.

Here's the biology bit:

1) <u>Haemoglobin</u> carries <u>oxygen</u> around your body from your lungs to your cells. Haemoglobin's the pigment which makes your blood <u>red</u>.

2) Haemoglobin <u>changes colour</u> depending on its <u>oxygen content</u>. When it's <u>rich</u> in absorbed oxygen it's <u>bright red</u> in colour (and it's called <u>oxyhaemoglobin</u>).

3) After giving up its oxygen to cells, the haemoglobin becomes <u>purply</u> in colour (and it's known as reduced haemoglobin).

And here's the physics bit…

1) A pulse oximeter has a <u>transmitter</u>, which emits two beams of <u>red light</u>. It also has a <u>photo detector</u> to measure light.

2) These are placed on either side of a <u>thin</u> part of the body, e.g. a <u>finger</u> or an <u>ear lobe</u>.

3) The beams of light pass through the tissue. On the way, some of the light is <u>absorbed</u> by the blood, reducing the amount of <u>light detected</u> by the detector.

4) The <u>amount of light absorption</u> depends on the <u>colour</u> of the blood. The colour of the blood depends on its <u>oxyhaemoglobin content</u>. In their arteries, healthy people normally have at least 95% oxyhaemoglobin and no more than 5% reduced haemoglobin.

5) <u>Reflection pulse oximetry</u> uses a similar technique — it reflects light off red blood cells instead of shining light through a part of the body.

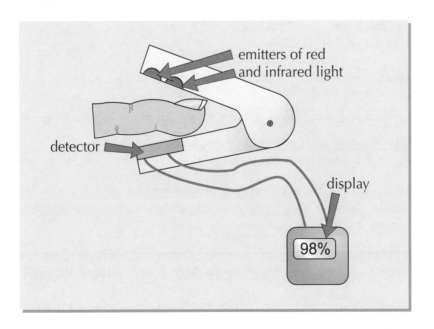

emitters of red and infrared light

detector

display

98%

Energy and Metabolic Rate

The body uses a lot of energy. Doctors tend to be interested in the rate it does it at.

Metabolic rate is the rate at which your body uses energy

1) The body is doing 'work' all the time. Some of this is obvious, e.g. the work you do when moving your muscles during exercise. Some of the energy transfers are less obvious, e.g. all the jobs your body does to keep you alive — digestion, tissue repair, your heart beating, your liver working, etc.

2) All of these mechanisms involve transferring energy (see page 121). And obviously, the more active you are, the more energy you're transferring.

Activity	kJ/min
Sleeping	4.5
Watching TV	7
Cycling (15 mph)	21
Jogging (5 mph)	40
Slow walking	14

3) The amount of energy used by the human body per second, minute, hour or day is called the metabolic rate.

4) The body gets its energy in the form of chemical energy from food. Most of this ends up as heat. (And kinetic energy if you move about. And sound energy too, if you move about and shout.)

Basal metabolic rate (BMR) is the rate you burn energy at rest

1) The rate at which the body transfers energy when it's at rest is called the basal metabolic rate (BMR). 'At rest' means doing nothing at all... not even digesting a meal.

2) The BMR tells you the minimum amount of energy needed to keep the body working properly (i.e. alive). Remember, the body continuously carries out loads of jobs which aren't immediately obvious.

3) To measure a person's BMR, they need to first go without food for a while (12 hrs, say) and then lie down in a room so that they're fully at rest. The amount of heat energy generated by their body is then measured over a certain time, and used to find the BMR.

BMR is usually measured as the energy (in kJ) needed per hour per m² of body surface area. Other useful measurements include joules per day, kilocalories per day, and even watts (joules per second).

Factors that can affect BMR include:

1) AGE: Children have a high basal metabolic rate because they need lots of energy to grow. The BMR reduces as you get older.

2) BODY FAT PERCENTAGE: The lower your body fat %, the higher your BMR. Muscle needs more energy than fat — even at rest. So BMR can also be affected by:

  • Gender — men generally have a lower % body fat than women — so a higher BMR.

  • Exercise — muscle-building and fat-burning exercise can lower body fat %, meaning a higher BMR as well as burning up calories during the process.

3) BODY SURFACE AREA: Higher surface area increases heat transfer, so this affects BMR too. For example, on a cold day, a larger body surface area loses heat quicker, meaning more energy is needed to keep the body warm, hence an increased BMR.

4) DIET: Starvation or sudden calorie reduction can dramatically reduce BMR. The body reacts to a sudden reduction in energy coming in by lowering the BMR so it needs less energy.

5) EXTERNAL TEMPERATURE: Exposure to cold makes the body increase its BMR to help keep warm.

Electricity and the Body

We've all seen <u>defibrillators</u> used in films and in TV medical dramas to 'jump-start' someone's heart. So it'd seem that electricity definitely has some kind of role in the body.

Muscle cells can generate potential differences

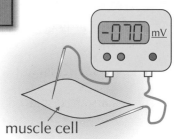

1) Between the <u>inside</u> of a muscle cell and the <u>outside</u>, there's a <u>potential difference</u> (a voltage). The potential difference across the <u>cell membrane</u> of a muscle cell at rest is called the <u>resting potential</u>.

2) These potential differences can be <u>measured</u> with really teeny tiny needle electrodes. The resting potential of a muscle cell is about <u>–70 mV</u> (millivolts).

muscle cell

> When a muscle cell is <u>stimulated</u> by an electrical signal, the <u>potential difference</u> changes from –70 mV to about <u>+40 mV</u>. This increased potential is called an <u>ACTION POTENTIAL</u>. The action potential passes down the length of the cell, making the muscle cell <u>contract</u>.

Electromyography (EMG) tests muscles

1) An EMG machine detects <u>small electrical signals</u> in muscles.

2) The results can be used to identify problems, e.g. <u>muscular dystrophy</u> and <u>motor neurone disease</u>.

3) EMG is also used as a <u>training tool</u> to help stroke victims learn to <u>use their muscles again</u>.

Action potentials in the heart chambers make the heart beat

1) The heart is a pump made of <u>muscle</u>, which is split up into <u>four chambers</u> — the <u>atria</u> at the top and the <u>ventricles</u> at the bottom.

2) When the heart <u>beats</u>, an <u>action potential</u> passes through the <u>atria</u>, making them <u>contract</u>. A fraction of a second later, <u>another</u> action potential passes through the <u>ventricles</u>, making them contract too.

3) Once the action potential has passed, the muscle <u>relaxes</u>.

4) These action potentials produce <u>weak electrical signals</u> on the skin.

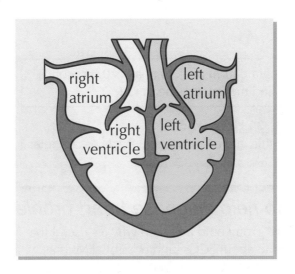

Electricity and the Body

Electrocardiographs can be used to monitor the <u>electrical activity</u> in the heart.

Electrocardiographs measure the action potentials of the heart

1) An <u>electrocardiograph</u> records the <u>action potentials</u> of the heart using electrodes stuck onto the chest, arms and legs.

2) For accurate readings the patient should lie or sit still and relax.

3) The results are displayed <u>on screen</u> or printed out as a <u>graph</u> called an <u>electrocardiogram</u> (ECG).

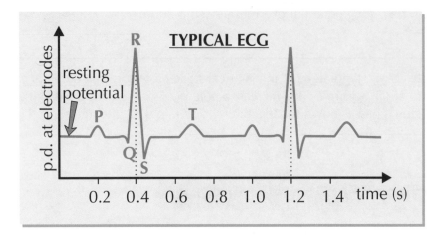

LEARN the <u>basic shape</u> and <u>what it means</u>:

- The horizontal line is just the <u>resting potential</u>.
- The 'blip' at <u>P</u> shows the <u>contraction</u> of the <u>atria</u>.
- The <u>QRS</u> blip shows the <u>contraction</u> of the <u>ventricles</u>. It's a <u>weird shape</u> because you've got the <u>relaxation</u> of the <u>atria</u> going on there too.
- And <u>T</u> shows the <u>relaxation</u> of the <u>ventricles</u>.

4) You can work out the <u>heart rate</u> from an ECG using this equation:

$$\text{frequency (hertz)} = \frac{1}{\text{time period (seconds)}}$$

Example

On the graph above, the time from peak to peak is 0.8 seconds.
Use this information to find the patient's heart rate.

Frequency = 1 ÷ 0.8 = <u>1.25 Hz</u>.
Multiply by 60 to convert this into a heart rate in beats per minute: 1.25 × 60 = <u>75 beats per min</u>.

ECGs can be used to help diagnose heart problems

Plenty of new stuff to learn here. You could be asked <u>why it's useful</u> that muscle cells generate a voltage — that's your cue to write about <u>ECGs</u> (and possibly EMGs as well). They could ask you to work out the heart rate based on an ECG — so don't forget the formula <u>f = 1/T</u>. Could be easy marks.

Warm-Up and Exam Questions

You know the score — make sure you can answer all of these questions before you move on...

Warm-Up Questions

1) What are the two main parts of an endoscope?
2) What is the device called that uses light to check oxygen level in the blood?
3) Give three examples of things the body uses energy for.
4) Where does the body get its energy from and what type of energy is it?
5) What do EMG and ECG stand for and what are they used for?
6) Write down the equation for calculating a heart rate from an ECG.

Exam Questions

1   Tamsin is told she has to have an operation on her gall bladder. The surgeon has told her she will be operated on using keyhole surgery.

   (a) Give an advantage of using keyhole surgery.
   (1 mark)

   (b) What instrument will the doctor use to see Tamsin's gall bladder during the operation?
   (1 mark)

   (c) During the operation, the surgeon uses a pulse oximeter to monitor Tamsin's blood.
        What part of the body would be suitable for placing in the detector of the oximeter?
   (1 mark)

2   Tony is a trainee doctor.
     The diagram shows an ECG
     trace he obtained from a patient.

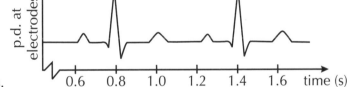

   (a) Tony says the ECG shape
        is caused by action potentials.

      (i)   What is an action potential?
      (1 mark)

      (ii) What happens when an action potential occurs in a muscle cell?
      (1 mark)

   (b) Work out the heart rate of Tony's patient in beats per minute.
   (2 marks)

   (c) What is happening to the heart when the time is 1.0 s?
   (1 mark)

3   Ab enjoys working out at the gym and wants to increase his basal metabolic rate (BMR).

   (a) What is meant by 'basal metabolic rate'?
   (1 mark)

   (b) Suggest a way Ab can increase his BMR.
   (1 mark)

Intensity of Radiation

Radiation — it's high time to get some facts straight on this subject, I think.
What is radiation, and how do you calculate its intensity — answers coming up...

Radiation is **energy emitted** from a **source**

The word 'radiation' is fairly common in Physics lessons and in everyday life.
It's often used to refer to ionising nuclear radiation (e.g. alpha, beta and gamma radiation)
— but 'radiation' actually covers a lot more than that.

Learn this really simple definition:

> **Radiation is energy emitted from a source**

This definition covers all types of radiation in Physics — e.g. light emitted from a star or alpha radiation spreading from a radioactive isotope. Radiation can be in the form of a wave (e.g. X-rays, visible light, infrared radiation) or a particle (e.g. alpha particles, positrons, neutrons).

Intensity of radiation = **rate of energy flow** to a **1 m² area**

1) The more intense the radiation, the more energy it carries. The more energy it carries, the more energy gets transferred when it hits an object.

2) Obviously, the surface area of the object affects the amount of radiation that smacks into it. A big object is going to catch more radiation than a small one.

3) Intensity is defined as the rate at which energy arrives at 1 square metre of surface. And 'rate of energy transfer' is POWER (page 127) — so the equation falls out as...

...intensity of radiation equals power of radiation divided by the area it falls on:

$$\text{intensity} = \frac{\text{power}}{\text{area}}$$

Example

The energy from a 100 W light bulb spreads over a surface of 4 m².
Calculate the intensity of radiation (falling on each m²).

Answer: I = P / A

$= 100 / 4$

$= \underline{25 \text{ W/m}^2}$ *The units are W/m², unsurprisingly.*

Intensity of Radiation

Intensity of radiation depends on distance from source

Think of a heater emitting <u>infrared radiation</u>. The <u>intensity of radiation decreases</u> as the <u>distance</u> from the source <u>increases</u>. You can tell that yourself, because the <u>further away</u> you are from the heater, the <u>less heat</u> energy you receive.

Here's why:

1) The intensity of the radiation depends on the area it's spread over.

2) If you move <u>twice</u> as far from the heater, the same radiation is being spread over <u>four times the area</u>.

3) That means you only receive $\frac{1}{2^2} = \frac{1}{4}$ of the heat energy.

4) This is known as an <u>inverse square</u> relationship.

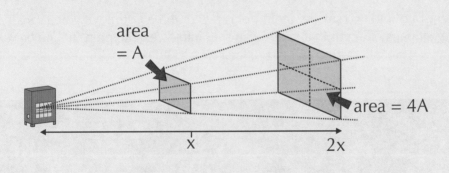

Intensity of radiation depends on what it's passing through

The <u>intensity of radiation</u> also depends on what it's <u>passing through</u>.

In practice, unless the radiation's passing through a <u>vacuum</u>, <u>some of the radiation</u> will always be <u>absorbed</u> along the way by the <u>medium</u> it's passing through — like air, for example.

Think about <u>sunglasses</u>. The dark glass <u>absorbs</u> some of the light, <u>reducing</u> its intensity.

Intense radiation transfers more energy when it meets an object

There are various units for radiation — don't learn them, but I'll mention them in case you see any in the exam. Hospital X-ray machines measure in <u>roentgens</u> — a unit of <u>intensity</u>. <u>Rads</u> are units of <u>radiation dose</u> equal to 0.01 J of energy per kg of body tissue. <u>Rems</u> and <u>sieverts</u> are units of <u>dose equivalent</u>, calculated by multiplying the dose by a 'quality factor' based on how <u>ionising</u> the radiation is.

Nuclear Bombardment

You can <u>bombard stable nuclei</u> with various different <u>particles</u> to make unstable nuclei. Depending on what you use, you get different products — a lot of which are used in <u>medicine</u> (which is why they're in this section).

The **splitting** of U-235 needs **neutrons**

1) <u>Uranium-235</u> is used in some <u>nuclear reactors</u> (and atomic <u>bombs</u>) (see page 184).

2) Uranium-235 (U-235) is actually quite <u>stable</u>, so it needs to be <u>made unstable</u> before it'll split — and this is done by firing slow-moving (<u>low-energy</u>) <u>neutrons</u> at the nucleus of a U-235 atom.

3) Only low-energy neutrons (called <u>thermal neutrons</u>) are <u>captured</u> by the nuclei — neutrons with higher energies just <u>bounce off</u> because they're travelling too fast.

4) When a thermal neutron joins the uranium nucleus it creates <u>U-236</u> — which is <u>really unstable</u>.

5) The U-236 then <u>splits</u> into two smaller atoms, plus two or three <u>fast-moving</u> neutrons. There are <u>different</u> pairs of atoms that U-236 can split into — e.g. krypton-90 and barium-144.

$$^{235}_{92}U + ^{1}_{0}n \rightarrow ^{90}_{36}Kr + ^{144}_{56}Ba + 2^{1}_{0}n + \text{energy}$$

These <u>fission products</u> usually have nuclei with a <u>larger proportion of neutrons</u> than atoms with a similar atomic number — this makes them <u>unstable</u> and <u>radioactive</u>.

Some of the products can be used for <u>medical applications</u>, e.g. <u>tracers</u> in medical diagnosis.

Nuclear Bombardment

Proton enrichment forms isotopes that emit positrons

1) Some radioactive isotopes are produced by bombarding stable elements with protons.

2) A proton is absorbed by the nucleus. This increases its proton number, so you get a new element.

3) A proton needs a lot of energy to be absorbed into the nucleus, so this process takes place in a circular particle accelerator called a cyclotron.

4) The radioisotopes made by proton enrichment are usually positron emitters. Remember, a positron is the antimatter version of an electron — same mass, opposite charge (see page 238).

5) Positron emitters are useful in hospitals — they're used in PET scanning. A PET scanner is a rather clever device used to monitor blood flow and metabolism — described in glorious detail on page 261.

6) Because some of these isotopes have short half-lives, it's important they're made close to where they will be used. For this reason, some hospitals have their own cyclotron to make the isotopes on-site.

Examples:

Here are three very useful radioisotopes made by proton bombardment, with equations showing how they're formed. These are all positron emitters and are all used in PET scanning.

1) Fluorine-18 is made using an isotope of oxygen called oxygen-18. Fluorine-18 has a half-life of just under 2 hours.

$$^{18}_{8}O + ^{1}_{1}p \longrightarrow ^{18}_{9}F + ^{1}_{0}n$$

2) Carbon-11 is made using nitrogen-14. Carbon-11 has a half-life of about 20 minutes.

$$^{14}_{7}N + ^{1}_{1}p \longrightarrow ^{11}_{6}C + ^{4}_{2}He$$

3) Nitrogen-13 is made using oxygen-16. It has a half-life of about 10 minutes.

$$^{16}_{8}O + ^{1}_{1}p \longrightarrow ^{13}_{7}N + ^{4}_{2}He$$

Here's a bit of extra practice to help you get familiar with this stuff:

Write a nuclear equation for the induced fission of uranium-235 into strontium-90 and xenon-144. Then cover the page and write an equation for the proton bombardment of ^{14}N to make a positron-emitting isotope and an alpha particle. (You'll find all the atomic numbers you need in the periodic table.)

Warm-Up and Exam Questions

Thought you'd escaped radiation in the last section? Not that easily my friend... Try these questions.

Warm-Up Questions

1) What is the definition of radiation?
2) What is the equation for working out the intensity of the radiation falling on a surface?
3) How is uranium-235 made unstable in a nuclear reactor?
4) What is a cyclotron and what is it used for?
5) Why do some hospitals have their own cyclotron?

Exam Questions

1   (a) A 60 W light bulb radiates all its energy evenly over a surface with an area of 3 m². Calculate the intensity of the radiation falling on the surface.

(1 mark)

       (b) The distance between the light bulb and the surface is now doubled. What is the intensity of the radiation falling on the surface now?

(2 marks)

       (c) Explain why, in practice, the radiation will be less intense than your answers to parts a) and b) suggest.

(2 marks)

2   Nuclear reactors often use uranium-235.

       (a) Explain how uranium-235 is split in a nuclear reactor.

(3 marks)

       (b) Write a nuclear equation for the fission of uranium-235 into barium-141 and krypton-92. Use the information from the table if necessary.

(3 marks)

Isotope	Symbol	Atomic no.
Uranium	U	92
Barium	Ba	56
Krypton	Kr	36

3   Fluorine-18 is a radioisotope made by proton enrichment.

       (a) What type of radiation does it emit?

(1 mark)

       (b) Write a nuclear equation for the production of fluorine-18 from oxygen-18. Use the information from the table if necessary.

Isotope	Symbol	Atomic no.
Fluorine	F	9
Oxygen	O	8

(2 marks)

       (c) Why are positron emitters useful in hospitals?

(1 mark)

Momentum Conservation

Ever wondered what happens to <u>momentum</u> when particles collide or get emitted from nuclei...

Momentum *is* **always** *conserved*

As you already found out on page 116, 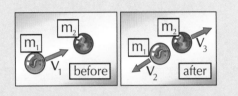 **momentum = mass × velocity**

1) In any collision or explosion, <u>momentum is always conserved</u> (as long as there aren't any external forces acting). This means that the <u>TOTAL MOMENTUM AFTER</u> is <u>EQUAL TO</u> the <u>TOTAL MOMENTUM BEFORE</u>.

2) The three situations that conservation of momentum might crop up in are:

Collision: bouncing off

This is similar to when a <u>fast-moving neutron</u> hits a nucleus and bounces off again.

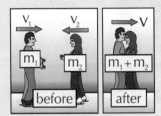

Collision: joining together

This example is similar to when a <u>neutron</u> or a <u>proton collides</u> with an atom and is <u>absorbed</u> into the nucleus.

Explosion: shot and recoil

This is like a particle being <u>emitted</u> from a nucleus. The nucleus <u>recoils</u> like a fired gun.

 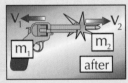

Example 1:

The diagram represents an iron nucleus capturing a neutron.

Find the final velocity.

  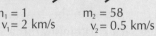

$m_1 = 1$, $v_1 = 2$ km/s; $m_2 = 58$, $v_2 = 0.5$ km/s; $m_1 + m_2 = 59$, $v_3 = ?$

momentum before = momentum after
$(1 × 2) + (58 × 0.5) = 59 v_3$
$31 = 59 v_3$
$v_3 = \underline{0.53\text{ km/s}}$ to the <u>right</u>

Momentum Conservation

You can never have too many examples, that's what I say. Here's another one for you.

Example 2:

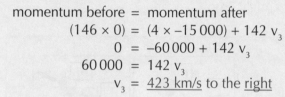

The diagram represents an alpha decay.

Find the velocity of the nucleus after the decay.

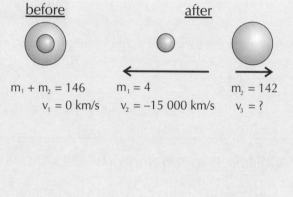

$m_1 + m_2 = 146$ $m_1 = 4$ $m_2 = 142$
$v_1 = 0$ km/s $v_2 = -15\,000$ km/s $v_3 = ?$

momentum before = momentum after
$$(146 \times 0) = (4 \times -15\,000) + 142\,v_3$$
$$0 = -60\,000 + 142\,v_3$$
$$60\,000 = 142\,v_3$$
$$v_3 = \underline{423 \text{ km/s to the } \underline{right}}$$

Positron/electron annihilation *is a good example*

Annihilation = complete destruction

1) When a particle meets its antiparticle (see page 241), the result is <u>annihilation</u>. <u>All the mass</u> of both particles is converted into <u>energy</u>, which is given off in the form of gamma rays.

2) When positrons and electrons meet, they tend to collide <u>head on</u> at the same speed and moving in <u>opposite directions</u> (don't ask why, you really don't want to know).

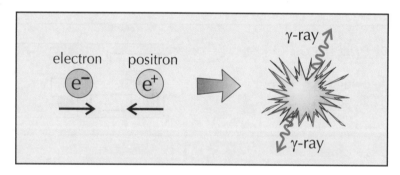

3) The particles have the <u>same mass</u> and <u>opposite velocities</u>, so the total momentum before the collision is <u>zero</u>.

4) Momentum is always conserved, so the gamma rays produced have a <u>total momentum of zero</u>.

5) The way that usually happens is that <u>two gamma rays</u> are produced which have the <u>same energy</u> but exactly <u>opposite velocities</u>.

6) <u>Mass/energy</u> is also conserved in this reaction — <u>all</u> the mass of the electron and positron has been converted into <u>energy</u>. It's Einstein's famous equation <u>$E = mc^2$</u> at work.

Matter + antimatter = massive KABOOM...

So hold on — you're telling me gamma rays have <u>momentum</u>... but I thought EM waves <u>didn't have any mass</u>... and if so, how can they have momentum... head hurts now. Bet they teach you the secret of that one at University. Something to do with Einstein and that mass/energy thing, no doubt.

Medical Uses of Radiation

Radiation can be incredibly <u>useful</u> in medicine, but any use of radiation carries some <u>risk</u>.

PET scanning involves positron/electron annihilation

PET is an abbreviation for <u>positron emission tomography</u>.   *Bit of a mouthful, you can see why they abbreviated it...*

It's a scanning technique used in hospitals to show tissue or organ <u>function</u>.
It's better than X-rays because X-rays just show <u>structure</u>. On the other hand, it's <u>extremely expensive</u>.

> 1) PET scans show areas of <u>damaged tissue</u> in the heart by detecting areas of <u>decreased blood flow</u>. This can reveal <u>coronary artery disease</u> and damaged or dead heart muscle caused by <u>heart attacks</u>.
>
> 2) PET scans can identify <u>active cancer tumours</u> by showing <u>metabolic activity</u> (see page 250) in tissue. <u>Cancer cells</u> have a <u>much higher metabolism</u> than healthy cells because they're growing like mad.
>
> 3) PET scans record <u>blood flow and activity</u> in the <u>brain</u>. This helps <u>research</u> and <u>treat</u> illnesses like Parkinson's, Alzheimer's, epilepsy, depression, etc.

And here's how it all works — get your brain in gear, because this is detailed:

1) Inject the patient with a substance used by the body, e.g. glucose, containing a <u>positron-emitting</u> radioactive isotope with a <u>short half-life</u>, e.g. ^{11}C, ^{13}N, ^{15}O or ^{18}F (see page 76). This radioactive substance is called a <u>radiotracer</u>.

2) Over an hour or so, the radiotracer <u>moves through the body</u> to the organs.

3) Positrons emitted by the radioisotope collide with <u>electrons</u> in the organs, causing them to annihilate (page 260), emitting <u>high-energy gamma rays</u>.

4) <u>Detectors</u> around the body record these gamma rays, and a computer builds up a <u>map</u> of <u>radioactivity</u> in the body.

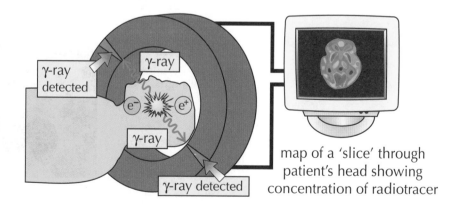

map of a 'slice' through patient's head showing concentration of radiotracer

5) The <u>distribution of radioactivity</u> matches up with <u>metabolic activity</u>. This is because <u>more</u> of the radioactive glucose (or whatever) injected into the patient is taken up and <u>used</u> by cells that are <u>doing more work</u> (cells with an <u>increased metabolism</u>, in other words).

Medical Uses of Radiation

PET scanning is a really clever technique.
But it's <u>not</u> the kind of thing that you want to be having done on a <u>weekly basis</u>.

Radiation exposure *should be* **limited**...

... for the following very good reasons:

> 1) Radiation can <u>destroy</u> a cell completely.
>
> 2) Radiation can <u>damage</u> a cell so that it can't divide.
>
> 3) Radiation can <u>alter the genetic material</u> in a cell. This can cause <u>mutations</u>. It can also make the cell divide and grow <u>uncontrollably</u> — this is <u>cancer</u>.

This is why it's important to <u>limit people's exposure</u> to radiation.

If you've ever had a scan that uses radiation, you may have noticed that the technician or doctor <u>leaves the room</u> while the scan is taken. This can seem a little worrying, but if they didn't they would be being exposed to radiation <u>every day</u> at work and that would obviously be a very bad idea.

> <u>NOTE</u>: <u>X-rays</u> and <u>PET scans</u> only use pretty <u>small doses</u> of radiation — 0.02 millisieverts (mSv) per chest X-ray, and 7 mSv per PET scan, compared to 2.2 mSv per year background radiation from just living in the UK.
>
> BUT... <u>ANY extra exposure</u> to radiation <u>increases</u> the <u>risk</u> of <u>cancer</u> — so it's recommended <u>not</u> to scan patients <u>too often</u> and only do it if <u>necessary</u>. It's all about <u>balancing risks</u>.

But sometimes **radiation treatment** *is still the* **best choice**

1) Radiation isn't just used for <u>diagnosing</u> problems — <u>radiotherapy</u> can be used to treat <u>cancer</u>.

2) <u>High-energy</u> X-rays or gamma rays are used to <u>destroy cancer cells</u>.

3) Radiation damages <u>cancer cells</u> much <u>more</u> than it damages <u>normal</u> cells.

4) Radiotherapy doesn't always lead to a <u>cure</u>, but it can often slow the progress of the illness or <u>reduce the suffering</u> when a patient is close to death.

5) Treatment that reduces suffering without curing an illness is called <u>palliative care</u>.

Doctors can use radiation to diagnose and treat diseases
PET scanning uses radiotracers to monitor blood flow and metabolic activity. And radiotherapy can be used to target and destroy cancer cells. But exposure to radiation always carries a risk.

Medical Research

New medical techniques provide tools for treating illness...

Physics examples:

1) <u>Endoscopes</u> for keyhole surgery.

2) <u>X-rays</u> to aid <u>diagnosis</u>.

3) <u>High-energy X-rays</u> to treat <u>cancer</u>.

4) <u>ECGs</u> to test heart function.

Biology examples:

1) <u>Gene therapy</u>.

2) Treatment for <u>infertility</u>.

3) <u>Embryo selection</u>.

4) Using <u>genetically modified animals</u> to produce <u>human enzymes</u> or even <u>body parts</u>.

Chemistry examples:

1) Research into <u>new drugs</u>.

2) New ways of <u>diagnosing illness</u> by analysing chemicals in the body.

You'll have come across plenty of these in earlier Science topics — more if you're doing GCSE Biology.

...but medical research brings ethical arguments with it

1) Just because medical research makes it <u>possible</u> to use a new technique, it doesn't mean we <u>should</u>.

2) The question of whether something is <u>morally</u> or <u>ethically</u> right or wrong <u>can't be answered</u> by more <u>experiments</u> — there is <u>no 'right' or 'wrong' answer</u>.

3) The best we can do is get a <u>consensus</u> from society — a <u>judgement</u> that <u>most people</u> are more or less happy to live by. <u>Science</u> can provide <u>more information</u> to help people make this judgement, and the judgement might <u>change</u> over time. But in the end it's up to <u>people</u> and their <u>consciences</u>.

New techniques have to be tested on people

1) New techniques are often tested first on cells grown in a lab, then on animals. Once they've been tested on animals and come out OK, they need to be <u>tested on people</u>.

2) There are <u>ethical arguments</u> that go with this:

- A new technique could have <u>harmful side effects</u>. Patients should be <u>aware</u> of possible side effects before they agree to take part in tests, but doctors <u>can't know for sure</u> what they could be.

- Lots of <u>ill patients</u> might want to get onto a medical trial. But places are limited.

- When a medical trial seems to show that a new technique <u>really works</u>, how long should it be before it's offered to <u>everyone</u>?

There are environmental and economic arguments too

Environmental factors:

<u>Unused medical radioisotopes</u> have to be <u>disposed of</u>. Some of them are <u>high-level nuclear waste</u> — radioisotopes used for radiation therapy have really long half-lives, and produce dangerous levels of radiation. The question of how (and where) to dispose of nuclear waste is a tricky one.

Economic factors:

<u>Companies</u> are unlikely to pay for research unless there's likely to be a <u>profit</u> in it — research into new drugs is funded by drugs companies that want to sell them at a profit.

Warm-Up and Exam Questions

You've already done momentum, but all this medical stuff's new, so get answering these questions...

Warm-Up Questions

1) Write down the equation for working out momentum.
2) Give three situations where you might use the conservation of momentum in calculations.
3) What is meant by 'annihilation' in physics?
4) What two characteristics are important for an isotope used in PET scanning?
5) What does 'consensus' mean and why is it relevant to medical physics?

Exam Questions

1   Rajesh has to go to hospital for a PET scan.

(a)   Give two examples of medical conditions that can be investigated by a PET scan.

(2 marks)

(b)   Rajesh sees that the radiology nurse is taking precautions to reduce her exposure to radiation. Give two ways in which radiation can harm the body.

(2 marks)

(c)   The table gives the type of radiation emitted by some radioactive isotopes.
It also gives their half-lives.

Isotope	Type of radiation	Half-life
A	alpha	300 years
B	beta	5 hours
C	gamma	6 years
D	positron	1 minute
E	positron	110 minutes

Which of the radioisotopes (A, B, C, D or E) would be most suitable for the following uses? Explain your answers.

(i)   To kill cancer cells by aiming beams of radiation at a tumour.

(2 marks)

(ii)   To be injected into the body as a medical tracer for use with a PET scanner.

(2 marks)

2   A positron and an electron, moving at exactly the same speed, collide, head on.

(a)   What is the total momentum before and after the collision?

(1 mark)

(b)   Each particle had mass and kinetic energy before the collision.
What happens to this mass and energy after the collision?

(2 marks)

(c)   What is this process called?

(1 mark)

3   A fast-moving neutron collides with a uranium-235 atom and bounces off. The diagram shows the particles before and after the collision. Find the velocity of the U-235 atom after the collision.

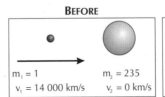

BEFORE

$m_1 = 1$ $m_2 = 235$
$v_1 = 14\ 000$ km/s $v_2 = 0$ km/s

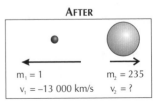

AFTER

$m_1 = 1$ $m_2 = 235$
$v_1 = -13\ 000$ km/s $v_2 = ?$

(3 marks)

Revision Summary for Section Eleven

Phew — that was the last topic in the book. But no rest for you yet — here's yet another shed-load of questions to help you find out what you don't know, so that you can go and revise it some more.

1) a)   What is total internal reflection?
   b)   When a light ray hits the inner boundary in an optical fibre, what happens if its angle of incidence is: i) less than the critical angle?   ii) more than the critical angle?
2) a)   What is an endoscope?
   b)   Describe how an endoscope works.
   c)   Name one medical technique made possible by endoscopy.
3) What colour is oxyhaemoglobin? What colour is reduced haemoglobin?
4) Describe how pulse oximetry works.
5) a)   What is metabolic rate?   b)   What is basal metabolic rate?
6) Give two examples of processes in the body that use energy even when you're completely at rest.
7) Who would you expect to have the lower BMR — a 48 year-old woman or a 29 year-old man?
8) What's meant by the 'resting potential' of a muscle cell?
9) What's meant by an 'action potential'?
10) An ECG shows that Karen's heart sends out a strong electrical signal every 0.7 seconds.
   a)*   Calculate Karen's heart rate in beats per minute.
   b)   As well as working out her heart rate, what could Karen's doctor use the ECG for?
11) a)   What's the proper definition of radiation?
   b)   Write down the equation for intensity of radiation.
12)*All the energy from a 1 kW electric bar heater radiates out evenly over a surface of 4 m². Calculate the intensity of the radiation at that surface.
13) What two factors affect the intensity of the radiation you receive from a 40 W light bulb?
14) What is a thermal neutron?
15)*Write a nuclear equation for the induced fission of uranium-235 into lanthanum-145 and bromine-88. (La has atomic number 57, Ba has atomic number 35.)
16)*Oxygen-18 is bombarded with protons to produce fluorine-18. Fluorine-18 is radioactive. What type of particles does it emit?
17)*A pink snooker ball travelling at 2 m/s hits a yellow snooker ball travelling at 1 m/s in the opposite direction. The pink ball stops. If both snooker balls have mass 150 g, what is the new velocity of the yellow ball?
18) a)   An electron and a positron travelling at equal speeds but in opposite directions annihilate when they collide, producing gamma rays. What is meant by the word 'annihilate'?
   b)   How is momentum conserved in this situation?
19) In a PET scan, what is 'PET' short for?
20) Why are the radioactive isotopes used in PET scans incorporated into substances like glucose?
21) PET scans can detect cancer tumours. Briefly explain how.
22) Give three ways in which radiation can harm cells.
23) What kind of radiation is used to treat cancer?
24) What is palliative care?
25) Scientists discover a new technique that uses radiation to treat cancer. What ethical arguments might there be about trialling this technique on human patients?

* Answers on page 281.

Answering Experiment Questions (i)

You'll definitely get some questions in the exam about <u>experiments</u>. They can be about any topic under the Sun — but if you learn the basics and throw in a bit of common sense, you'll be fine.

Read the question carefully

The question might describe an <u>experiment</u>, e.g. —

Roger had a plastic toy diver and three different sized plastic parachutes. He investigated which parachute was the most effective in slowing down the diver when falling through water.

Toy diver

Small parachute

Medium parachute

Large parachute

Roger timed the diver's fall to the bottom of a fish tank (without parachute). He did this three times. He then attached the smallest parachute to the toy and timed its fall, again three times. He did the same for the medium and the large parachutes.

1. What is the independent variable?
   The size of the parachute

2. What is the dependent variable?
   The time taken to fall

3. Give two variables that must be kept the same to make it a fair test.
   1. The distance the toy had to fall.
   2. The mass of the toy (ideally the same toy would be used)

4. What is the control in this investigation?
   The toy with no parachute

The <u>independent variable</u> is the thing that is <u>changed</u>.

The <u>dependent variable</u> is the thing that's <u>measured</u>.

To make it a <u>fair test</u> you've got to keep <u>all</u> other variables the same, or you won't know if the <u>only thing</u> affecting the dependent variable is the <u>independent variable</u>.

There are <u>loads</u> of other things that must be kept the same throughout this experiment. You could also have put the <u>shape of the parachute</u> or the <u>size of tank</u>, etc.

It's easy to keep the variables the same in this experiment as it's in a <u>laboratory</u>. But it can sometimes be <u>trickier</u>. E.g. if you were doing this in a <u>pond</u>, it'd be hard to ensure that you had exactly the <u>same conditions</u> each time — the <u>depth</u> may vary, the <u>water</u> could be <u>moving</u>, etc.

It's even harder to make investigations involving <u>people</u> fair.

If, say, the effect of a person's age on their blood pressure was being investigated, there'd be <u>loads of other variables</u> to consider — weight, diet and whether someone's a smoker could make a <u>big difference</u> to their blood pressure.

To make it a <u>fairer</u> test, it would be better if just nonsmokers with a similar weight and diet were used.

A <u>control</u> isn't really part of the experiment, but it uses the <u>same conditions</u> as the rest of the experiment. You can compare the results of the experiment with those of the control group, and see if the same things might have happened <u>anyway</u>. Control groups make results <u>more meaningful</u>.

In this experiment the toy might <u>fall slower</u> without any parachute — unlikely, but with a control group you can check for this.

Control groups are used when <u>testing drugs</u>. People can feel better just because they've been given a drug that they <u>believe</u> will work. To rule this out, researchers give one group of patients <u>dummy pills</u> (called placebos) — but they <u>don't tell them</u> that their pills aren't the real thing. This is the control group. By doing this, they can tell if the real drug is actually working.

Answering Experiment Questions (ii)

5. Why did Roger time each fall three times, instead of just once?

To check for anomalous results and make the results more reliable.

Sometimes unusual results are produced — repeating an experiment gives you a better idea what the correct result should be.

6. The table below shows the times taken by the diver to fall to the bottom of the tank.

parachute size ↓	Time for Fall 1 (s)	Time for Fall 2 (s)	Time for Fall 3 (s)	Mean Time (s)
none	1.2	1.5	1.2	
small	2.1	2.4	2.4	2.3
medium	2.8	2.6	2.5	2.6
large	3.2	3.4	(5.1)	3.3

When an experiment is repeated, the results will usually be slightly different each time.

The mean (or average) of the measurements is usually used to represent the values.

The more times the experiment is repeated the more accurate and reliable the average will be.

To find the mean:

Add together all the data values and divide by the total number of values in the sample.

The range is how far the data spreads.

You just work out the difference between the highest and lowest numbers.

a) Calculate the mean time taken for the toy to fall without a parachute.
Mean = (1.2 + 1.5 + 1.2) ÷ 3 = 1.3 s

b) What is the range of the times for the fall without a parachute?
_1.5 – 1.2 = 0.3 s_

7. One of the results in the table is anomalous. Circle the result and suggest why it may have occurred.
The parachute may have had an air bubble trapped inside it.

If one of the results doesn't seem to fit in, it's called an anomalous result. You should usually ignore an anomalous result. It's been ignored when the mean was worked out.

This is a random error — it only happens occasionally.

If the same mistake is made every time, it's a systematic error, e.g. if you were looking at the tank from too high up, it might look like the toy hit the bottom before it actually did. So you'd stop the stopwatch too soon, meaning all your timings would be a bit quick.

8. What conclusion can you draw from these results?
Parachutes slow down the fall of the toy, and the bigger the parachute (made from the same material) the slower the toy falls.

Be careful that your conclusions match the data you've got and don't go any further than your results actually show.

For example, you can't say that the bigger the parachute the slower the fall, because the results may be totally different with another type of material, e.g. lace.

These experiments are just examples — don't learn the details

The point of these pages isn't to remember the details of the experiments — they could ask these questions about any random experiment. You need to know what your different kinds of variable are, what's the control, how to make it a fair test and what to do with your results when you've got them.

Answering Experiment Questions (iii)

Use *sensible measurements* for your *variables*

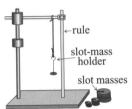

←rule

slot-mass
holder

slot masses

Jamie did an experiment to see how the length of a wire changed depending on the weight hanging on it.

He started off with just a slot-mass holder hanging on the wire and made sure that the starting length was 10 cm for each wire. He then gradually added slot masses until the wire broke. For each mass he added, he measured the new length. He did the experiment using iron, steel, aluminium, copper and brass wires.

Before he started, he did a trial run, which showed that most wires snapped under loads of between 10 kg and 30 kg and extended to lengths of between 2 cm and 4 cm.

1. What kind of variable was the list of metals?

   A A continuous variable ☐

   B A categoric variable ✓

   C An ordered variable ☐

   D A discrete variable ☐

Categoric variables are variables that can't be related to size or quantity — they're <u>types</u> of things. E.g. <u>names of metals</u> or <u>types of fertiliser</u>.

<u>Continuous data</u> is <u>numerical data</u> that can have <u>any value</u> within a range — e.g. length, volume, temperature and time.

Note: You <u>can't</u> measure the <u>exact value</u> of continuous data. Say you measure a height as 5.6 cm to the nearest mm. It's not <u>exact</u> — you get a more precise value if you measure to the nearest 0.1 mm or 0.01 mm, etc.

<u>Ordered variables</u> are things like <u>small, medium and large lumps</u>, or <u>warm, very warm and hot</u>.

<u>Discrete data</u> is the type that can be counted in chunks, where there's no in-between value, e.g. <u>number of people</u> is discrete not continuous because you can't have half a person.

2. Jamie should add masses in intervals of...

   A 0.5 g. ☐

   B 10 g. ☐

   C 1 kg. ✓

   D 10 kg. ☐

It's important to use <u>sensible values</u> for variables.

It's no good using <u>really heavy</u> slot masses as the wire might snap immediately. But on the other hand, adding <u>really weedy ones</u> like 0.5 g one at a time could take ages.

3. The rule used to measure the wire length should be capable of measuring...

   A to the nearest mm. ✓

   B to the nearest inch. ☐

   C to the nearest 10 cm. ☐

   D to the nearest cm. ☐

A rule measuring only to the nearest cm, or bigger, would <u>not</u> be <u>sensitive enough</u> — the changes in length are likely to be quite small, so you'd need to measure to the <u>nearest mm</u> to get the <u>most precise</u> results.

The <u>sensitivity</u> of an instrument is the <u>smallest change</u> it can detect, e.g. some balances measure to the nearest <u>gram</u>, but really sensitive ones measure to the nearest <u>hundredth of a gram</u>. For measuring <u>tiny changes</u> — like from 2.00 g to 1.92 g — a sensitive balance is needed.

You also have to think about the <u>precision</u> and <u>accuracy</u> of your results.

Measurements (of the same thing) that are very <u>precise</u> will be close together. Really <u>accurate</u> measurements are those that have an <u>average value</u> that's <u>really close</u> to the <u>true answer</u>. So it's possible for results to be precise but not very accurate, e.g. a fancy piece of lab equipment might give results that are precise, but if it's not calibrated properly those results won't be accurate.

Answering Experiment Questions (iv)

Once you've collected all your data together, you need to <u>analyse</u> it to find any <u>relationships</u> between the variables. The easiest way to do this is to draw a graph, then describe what you see...

Graphs are used to show *relationships*

These are the results Jamie obtained with the brass wire.

Load (kg)	0	1	2	3	4	5	6	7	8	9	10
Extension (mm)	0	1	2	2	3	4	5	5	7	8	9

4. a) Nine of the points are plotted below. Plot the remaining **two** points on the graph.

To plot the points, use a <u>sharp</u> pencil and make a <u>neat</u> little cross.

 nice clear mark

 smudged unclear marks

b) Draw a straight line of best fit for the points.

A line of best fit is drawn so that it's easy to see the <u>relationship</u> between the variables. You can then use it to <u>estimate</u> other values.

Scattergram to show the extension of a brass wire with different amounts of load

Extension (mm) vs Load (kg)

anomalous result

When drawing a line of best fit, try to draw the line through or as near to as many points as possible, ignoring <u>anomalous</u> results. In this case, it's also got to go through the <u>origin</u> (0, 0) as you know there'd be no extension without any load.

This is a <u>scattergram</u> — they're used to see if two variables are <u>related</u>.

This graph shows a <u>positive correlation</u> between the variables. This means that as one variable <u>increases</u>, so does the other one.

5. Estimate the load you would need if you wanted the brass wire to extend by 4 mm.

Estimate of load = ...4.5 kg (see graph)......

6. What can you conclude from these results?

There is a positive correlation between the load and the extension. Each additional mass causes the wire to extend further.

The other correlations you could get are:

 <u>Negative correlation</u> — this is where one variable <u>increases</u> as the other one <u>decreases</u>.

 <u>No correlation</u> — this is where there's <u>no obvious relationship</u> between the variables.

In lab-based experiments like this, you can say that one variable <u>causes</u> the other to change. The extra load <u>causes</u> the wire to extend further. You can say this because everything else has <u>stayed the same</u> — nothing else could be causing the change.

Page 12

Warm-Up Questions

1) A type of energy / a measure of energy.

2) When there is a temperature difference between two places.

3) Breaking the bonds between molecules to turn the liquid into a gas.

4) The amount of energy needed to melt 1 kg of a substance without changing its temperature.

Exam Questions

1   B *(1 mark)*
18 000 (energy needed) ÷ 80 (temperature change) ÷ 0.5 (mass of iron)
= 450 J/kg/°C

2   A — 3 *(1 mark)*
B — 2 *(1 mark)*
C — 4 *(1 mark)*
D — 1 *(1 mark)*

3   (a) 2000 × 60 × 2 = 240 000 J *(2 marks, allow 1 mark for correct working)*

(b) Mass = energy ÷ specific latent heat
= 240 000 ÷ 2 260 000
= 0.106 kg (= 106 g)
(2 marks, allow 1 mark for correct working)

Page 18

Warm-Up Questions

1) Particles that vibrate faster than others pass on their extra kinetic energy to their neighbours.

2) Heated air expands, so it becomes less dense than the surrounding cooler air and rises.

3) infrared (radiation)

4) E.g. make the surface darker in colour, make the surface rougher.

5) Any three of, e.g. loft insulation / cavity wall insulation / draught proofing / double glazing / using thick curtains.

Exam Questions

1   C *(1 mark)*

2   D *(1 mark)*
Light and shiny surfaces are best for reflecting heat radiation.

3   (a) 300 × 0.25 = £75 *(1 mark)*

(b) 300 − 255 = £45 saved per year *(1 mark)*.
Payback time = cost ÷ saving per year = 350 ÷ 45 = 7.8 years *(1 mark)*.

Page 23

Warm-Up Questions

1) chemical energy

2) Energy cannot be created or destroyed, only converted from one form to another.

3) More of the input energy is transformed into useful energy in modern appliances.

4) E.g. wind-up radio, clockwork toy.

5) Some energy is always wasted and so all the input energy isn't transformed usefully.

Exam Questions

1   B *(1 mark)*
Efficiency = Useful energy output ÷ Total energy input.

2   A *(1 mark)*
Each square represents 10 J. The bulb is 15% efficient, so when 100 J of energy are input, the useful output must be 1.5 squares wide.

3   (a) 1200 − 20 − 100 = 1080 J *(1 mark)*
Energy cannot be destroyed or created, so the total energy output must equal the energy input.

(b) By reducing the amount of energy wasted as sound / by having a quieter motor *(1 mark)*.
Only the energy converted to sound is wasted — kinetic and heat energy are ... you want from a hairdrier.

Pages 34-35

Warm-Up Questions

1) Any two of, e.g. it releases greenhouse gases/contributes to global warming / it causes acid rain / coal mining damages the landscape.

2) Nuclear power stations produce radioactive waste, which is dangerous and difficult to dispose of / risk of a catastrophic accident.

3) Organic matter that can be burnt to release energy.

4) Pumped storage is a method of storing electricity whereas hydroelectric power schemes actually generate electricity.

5) Any two of, e.g. they're expensive to install / are inefficient / can only generate electricity when there is enough sunlight.

Exam Questions

1   (a) Heat energy from inside the Earth *(1 mark)*.

(b) The source of energy will never run out *(1 mark)*.

(c) Because there are no hot rocks near the surface of the Earth in the U.K. *(1 mark)*

2   D *(1 mark)*
Tides are caused by the pull of the Sun and Moon's gravity.

3   A — 3 *(1 mark)*
B — 4 *(1 mark)*
C — 1 *(1 mark)*
D — 2 *(1 mark)*

4   (a) 2 000 000 ÷ 4000 = 500 *(1 mark)*

(b) If the wind isn't blowing strongly, the turbines will not generate as much as 4000 W each *(1 mark)*.

(c) Any two of, e.g. they believe it would spoil the view (visual pollution) / cause noise pollution / kill or disturb local wildlife *(1 mark each)*.

5   A *(1 mark)*
Wind, solar and wave power all depend on the weather, which is changeable.

6   (a) (i) by (thermal) radiation *(1 mark)*

(ii) by conduction through the metal pipe *(1 mark)*

(iii) by convection currents in the water *(1 mark)*

(b) Because black surfaces are good absorbers of heat radiation *(1 mark)*.

7   (a) Any two of, e.g. wave / tidal / geothermal / biomass *(1 mark for each)*.

(b) Any two of, e.g. set-up time / set-up costs / running costs / impact on environment / social impact *(1 mark for each)*.

Page 36

Revision Summary for Section One

2)   900 J/kg/°C

4)   791 kJ

10) a) 2 years,

b) light bulb B

14) a) 80 J

b) 20 J

c) 80 ÷ 100 = 80%

17)   0.7 (or 70%)

Pages 44-45

Warm-Up Questions

1) The flow of electrons/charge round a circuit.

2) Alternating current (AC) changes direction whereas direct current (DC) always flows in the same direction.

3) It increases.

4) The creation of a voltage (and maybe current) in a wire which is experiencing a change in magnetic field.

5) The network of cables and pylons that distributes electricity across the country.

6) voltage = current × resistance / V=IR / resistance = voltage ÷ current / current = voltage ÷ resistance

Exam Questions

1   A *(1 mark)*

2   A — 4   *(1 mark)*

     B — 1   *(1 mark)*

     C — 2   *(1 mark)*

     D — 3   *(1 mark)*

3   D *(1 mark)*

4 (a) Because electrical current is generated. *(1 mark)*

   (b) Any one of, e.g. move the magnet out of the coil / move the coil away from the magnet / insert the south pole of the magnet into the same end of the coil / insert the south pole of the magnet into the other end of the coil. *(1 mark)*

   (c) Any one of, e.g. push the magnet into the coil more quickly / use a stronger magnet / add more turns to the coil. *(1 mark)*

   (d) Zero / no reading. *(1 mark)* *Movement is needed to generate a current.*

5 (a) Chemical energy *(1 mark)*

   (b) (i) 25 000 × 1000 = 25 000 000 kJ = 25 000 MJ *(1 mark)*

      (ii) (10 000 ÷ 25 000) × 100 = 40% *(1 mark)*

Page 47

Top Tip

1) Power = voltage × current = 230 × 12 = 2760 W = 2.76 kW

2) Time = energy ÷ power = 0.5 ÷ 2.76 = 0.181 h = 11 min

Page 48

Warm-Up Questions

1) The rate of transfer of electrical energy.

2) kilowatt hours (kWh)

3) power = current × voltage

4) units of energy (kWh) = power (kW) × time (hours)
= 3.1 × (1.5 ÷ 60) = 0.0775 kWh

Exam Questions

1   C *(1 mark)* $I = P \div V = 1200 \div 230 = 5.22\,A$

2   B *(1 mark)*

3 (a) 8.75 kWh = 8.75 × 1000 × 60 × 60 = **31 500 000 J** *(1 mark)*

   (b) 8.75 × 14.4 = **126p** (=£1.26) *(1 mark)*

   (c) £1.26 × 90 = **£113.40** *(1 mark)*

   (d) 12 × 8.75 = 105p, 126 − 105 = **21p** *(1 mark)*
     (OR 14.4 − 12 = 2.4, 2.4 × 8.75 = 21p)

Pages 55-56

Warm-Up Questions

1) A wave enters a different medium (with its wavefronts at an angle to the boundary) and changes direction. This happens because the wave changes speed in the new medium.

2) The number of complete wavelengths that pass a point within 1 second.

3) Interference occurs when waves with similar frequencies combine, so that both signals are distorted.

4) Diffraction

5) When light is travelling from a denser medium towards a less dense medium and its angle of incidence at the boundary is greater than the critical angle.

6) A thin glass or plastic fibre which is used to transmit light/IR waves. Because the light no longer hits the boundary within the optical fibre at a big enough angle, and so is not internally reflected (and transmitted along the fibre).

Exam Questions

1   C *(1 mark)*

2   A *(1 mark)* $\lambda = v \div f = 300\,000\,000 \div 100\,000\,000 = 3\,m.$
Don't forget that you have to convert MHz to Hz before doing the calculation.

3 (a) 5 cm *(1 mark)*

   (b) 1 complete wave would pass a point every 2 seconds, so f = 1,
     2 = **0.5 Hz** *(1 mark)*

   (c) It will halve. *(1 mark)*
     Because frequency and wavelength are inversely proportional.

4   A — 2 *(1 mark)*

     B — 3 *(1 mark)*

     C — 4 *(1 mark)*

     D — 1 *(1 mark)*

5 (a)

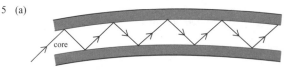

    (1 mark)

   (b) The light ray meets the core/outer boundary at an angle greater than the critical angle, so is totally internally reflected, and this happens repeatedly. *(1 mark)*

Page 59

Warm-Up Questions

1) It ionises atoms/molecules within cells — this can cause e.g. mutations in DNA, or kill cells.

2) It causes heating / can cause burns.

3) Because they damage / cause thinning of the ozone layer (which protects us from UV radiation).

4) Ultraviolet

Exam Questions

1   C *(1 mark)*
     No one knows for sure whether microwave radiation from phones and masts is causing any health problems.

2   A — 3 *(1 mark)*

     B — 4 *(1 mark)*

     C — 2 *(1 mark)*

     D — 1 *(1 mark)*

3 (a) (i) Hazard: is absorbed by water molecules, causing heating which could damage living cells. *(1 mark)* Use: any one of, e.g. mobile phone signals/communication / cooking *(1 mark)*

      (ii) Hazard: any one of, e.g. it's highly ionising / it can cause cell damage/mutations in DNA / it can cause cancer / it can cause radiation sickness. *(1 mark)* Use: any one of, e.g. sterilising food/medical instruments / treating cancer / as a medical or industrial tracer. *(1 mark)*

   (b) Any one of, e.g. stay in the shade / wear sunscreen / keep covered up with clothing. *(1 mark)*

Page 64

Warm-Up Questions

1) Any two of, e.g. because they are not absorbed by the Earth's atmosphere / they are not dangerous / they can be sent long distances.

2) Certain frequencies of microwave are absorbed by water molecules (which are present in all food). This heats the water and so heats the food.

3) Different wavelength/frequency microwaves are used for each purpose.

4) (short wavelength) radio waves

5) The laser is reflected by the surface of the CD into a light sensor. The laser is reflected from the pits and lands slightly differently. The light sensor monitors these changes and converts them into an electrical signal.

Exam Questions

1   C *(1 mark)*

2 (a) It acts like a camera, recording light reflected from the iris. (This image is then processed by computer into a code.) *(1 mark)*

   (b) Any one of, e.g. iris scanning is quick and easy / iris patterns don't normally change during a person's lifetime / there is very little chance of mistaking one iris for another. *(1 mark)*

   (c) Any one of, e.g. eye injuries or surgery can occasionally change the iris pattern / people feel that their personal freedom may be being threatened / iris pattern data is believed to be impossible to fake and so genuine identity theft or mistakes may not be believed. *(1 mark)*

3 (a) Because X-rays mostly pass straight through soft tissue, so the structure of soft tissues doesn't appear on X-ray images. *(1 mark)*

(b) E.g. leave the room while the X-ray image is being taken, wear a lead apron (if they need to remain in the room). *(2 marks)*

4   Advantage: any one of, e.g. doctors can check on the baby's development / parents can find out the sex of the baby. *(1 mark)*

Disadvantage: any one of, e.g. scientists are not certain that ultrasound is completely safe / parents may be more likely to want an abortion if their baby is found to be the "wrong sex" or disabled. *(1 mark)*

Page 69

Warm-Up Questions

1) Digital signals can only take two values, on and off, whereas analogue signals can take any values within a given range.

2) P-waves

3) P-waves

4) They are refracted as the properties of the Earth's interior change.

5) Any one of, monitoring ground water levels / detecting foreshocks / monitoring radon gas emissions / monitoring animal behaviour.

Exam Questions

1   D *(1 mark)*

2   A *(1 mark)*

3 (a) on or off (1 or 0) *(1 mark)*

(b) Both types of signal will have lost energy and weakened. *(1 mark)*

(c) Interference on digital signals can be much more easily filtered out by his radio. *(1 mark)*

Page 70

Revision Summary for Section Two

4)   4.8 ohms

11) a) Units = power × time = 2.5 × (2/60) = 0.0833 = 0.08 kWh

b) 0.0833 × 12 = 1 p

16)   150 m/s

20)   A and D (which are identical)

Page 73

Top Tip

beta

Page 78

Warm-Up Questions

1) An unstable atom emits radiation and becomes more stable, often changing into a different element.

2) Atoms which have the same number of protons but different numbers of neutrons.

3) Alpha, beta and gamma

4) Gamma radiation

5) Any one of, e.g. cosmic rays, rocks, radon gas, food, building materials, human activity (e.g. fallout from nuclear bomb tests), living things.

Exam Questions

1   A — 4 *(1 mark)*

B — 1 *(1 mark)*

C — 3 *(1 mark)*

D — 2 *(1 mark)*

2   C *(1 mark)*

3 (a) The time taken for half of the unstable nuclei in a sample to decay / the time taken for the count rate to halve. *(1 mark)*

(b) one quarter / 25% *(1 mark)*

(c) beta *(1 mark)*

Page 85

Warm-Up Questions

1) A weak alpha source is used to ionise the air between two electrodes so that a current can flow. If the alpha radiation is absorbed by smoke, the current stops and the alarm sounds.

2) Because it is ionising and can damage cells.

3) Any two of, e.g. never look directly at the source / always handle a source with tongs / never allow the source to touch the skin / never have the source out of its lead-lined box for longer than necessary.

4) Any one of, e.g. wear lead-lined suits / work behind lead/concrete barriers / use robotic arms to do tasks in highly radioactive areas.

5) Any one of, e.g. medical tracers / treatment of cancer.

6) The level of C-14 in the atmosphere has always been constant (though in practice variations are taken into account). / All living things take in the same proportion of their carbon as C-14. / Substances haven't been contaminated by a more recent source of carbon.

Exam Questions

1   A *(1 mark)*

2   D *(1 mark)*

3 (a) Beta radiation is directed through the paper towards a detector *(1 mark)*. If the paper's thickness changes, so does the amount of radiation reaching the detector, and a control moves the paper rollers accordingly. *(1 mark)*

(b) Alpha radiation would be completely stopped by paper. *(1 mark)*

(c) Gamma radiation would not be blocked by even thick paper and so couldn't provide any information about thickness. *(1 mark)*

Page 91

Warm-Up Questions

1) Between Mars and Jupiter.

2) dust, gas and ice

3) Any one of, e.g. asteroids orbit in the same plane as the planets; comets do not / asteroids were formed at the same time as the planets; comets were not / comets have highly elliptical orbits; asteroids do not.

4) A very large group of stars/solar systems.

5) Because its gravity is so strong that light can't escape.

Exam Questions

1   A — 3

B — 4

C — 2

D — 1 *(1 mark for each correct answer)*
1 must be the comet as it orbits in a different plane from the planets.

2   C *(1 mark)*

3 (a) Explosions on the Sun's surface that throw out huge amounts of energy and charged particles/cosmic rays. *(1 mark)*

(b) (i) The Earth is shielded by its magnetic field. *(1 mark)*

(ii) The Aurora Borealis / northern lights. *(1 mark)*
Cosmic rays pass on energy to particles in the atmosphere which then emit visible light.

(c) (i) Any one of, e.g. navigation, spying, weather forecasting, communications. *(1 mark)*

(ii) The charged particles can cause a surge of current which may damage satellites' electrical components. *(1 mark)*

Page 95

Warm-Up Questions

1) A very large/heavy star.

2) The low frequency microwave radiation that comes from all parts of the Universe.

3) A red giant.

4) It may contract and end in a 'Big Crunch' or it may expand forever.

5) Our Sun (and Solar System) formed from material created in the supernova of a previous star.

Exam Questions

D *(1 mark)*

A *(1 mark)*

(a) All the matter and energy was in a very small space, then there was an explosion/a 'Big Bang'. *(1 mark)*

(b) The further the galaxy, the greater the red shift *(1 mark)*. This shows that the more distant the galaxy, the faster it's moving away from us. This must mean that the Universe is expanding. *(1 mark)*

(c) It depends on what the total mass of the Universe is, but we don't know how much 'dark matter' there is. *(1 mark)*

Page 100

Warm-Up Questions

- To minimise light pollution and pollution from dust, etc.

- Because light pollution reduces the clarity of images of space.

- Any one of, e.g. they don't have to carry food/water/oxygen / they don't need to be protected from the conditions that would be lethal to humans / they don't have to return.

- It increases the resolution of the images produced. / It allows fainter objects to be 'seen'.

- The level of detail the images include.

Exam Questions

(a) Because there would be a delay of several minutes between one person sending a message and the other receiving it. *(1 mark)*

(b) Any two of, e.g. food / water / oxygen. *(1 mark for two correct answers)*

(c) Any two of, e.g. muscle wastage / loss of bone tissue / mental stress. *(1 mark for each correct answer)*

(d) Any one of, e.g. the astronauts can carry out repairs / people can think for themselves to overcome problems. *(1 mark)*

(a) The Earth's atmosphere will not interfere with the view of a space telescope. *(1 mark)*

(b) With a lighter 'payload', less fuel was needed to take off from Earth and put the telescope into orbit. This made it cheaper. *(1 mark)*

(a) Linking up lots of smaller ones. *(1 mark)*

(b) To get good resolution the dish has to be large compared to the wavelength *(1 mark)* and radio waves have a longer wavelength than light waves *(1 mark)*.

(c) Because X-rays are absorbed by the Earth's atmosphere. *(1 mark)*

Page 101

Revision Summary for Section Three

3) 3 half-lives = 17 190 years

Pages 107-108

Warm-Up Questions

) 10 m ÷ 5 s = 2 m/s

) Acceleration — m/s², mass — kilograms (kg), weight — newtons (N)

) Speed

) (30 – 0) ÷ 6 = 30 ÷ 6 = 5 m/s²

) Gravity

Exam Questions

(a) Because its direction is constantly changing. *(1 mark)*

(b) Time = distance ÷ speed = 2400 ÷ 45 = 53.3 s (to 1 d.p) *(2 marks, allow 1 mark for correct working)*

(c) Acceleration = change in velocity ÷ time taken = (59 – 45) ÷ 5 = 14 ÷ 5 = 2.8 m/s² *(2 marks, allow 1 mark for correct working)*

(a) (i) 200 m *(1 mark)* Read the distance travelled from the graph.

(ii) 200 ÷ 15 = 13.3 m/s (to 1 d.p.) *(2 marks, allow 1 mark for correct working)*

(b) 13 s (allow answers from 11 s to 15 s) *(1 mark)* The bus is stationary between about 33 and 46 seconds.

(c) The bus is travelling at constant speed (10 m/s) back towards the point it started from. *(1 mark)*

(d)

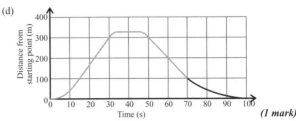

(1 mark)

3 (a) The cyclist is travelling at a constant velocity of 3 m/s. *(1 mark)*

(b) The cyclist's speed is constantly decreasing. *(1 mark)*

(c) (3 – 0) × (5 – 2) ÷ 2 = 4.5 m *(1 mark)* Remember, distance travelled is the area under the graph.

4 (a) Acceleration due to gravity, g, is lower on Mars than on Earth *(1 mark)*, because Mars has a smaller mass than Earth. *(1 mark)* So the ball's weight is less on Mars than on Earth. *(1 mark)*

(b) 3 ÷ 1.1 = 2.7 (to 1.d.p) So, the spring extends 2.7 times as far on Earth, so g on Earth must be 2.7 times bigger than on Mars. *(1 mark)* 10 ÷ 2.7 = 3.7 N/kg (to 1.d.p) *(1 mark)*

5 (a) Acceleration = change in velocity ÷ time taken = (10 – 0) ÷ 1 = 10 m/s² *(1 mark)*

(b) Change in velocity = acceleration × time = 10 × 3 = 30 m/s *(1 mark)*

(c) Assume g = 10 m/s². Weight = mass × g = 0.12 × 10 = 1.2 N *(1 mark)*

(d) None — the stone's acceleration due to gravity is determined by the mass of the Earth, not the stone. *(1 mark)*

6 All of them. *(1 mark)* Everything with mass exerts a gravitational force.

Page 114

Warm-Up Questions

1) 0 N

2) The acceleration also doubles.

3) 3 + 3 – 1 = 5 N towards the shore

4) Against motion.

Exam Questions

1 (a) (i) 900 N – 900 N = 0 N *(1 mark)*

(ii) The parachutist could be falling at a constant (terminal) velocity or could be at rest (e.g. stood on the ground). *(1 mark)*

(b) (i) Terminal velocity is reached when the force of air resistance equals the parachutist's weight. *(1 mark)* Weight = mass × g = 70 × 10 = 700 N *(1 mark)*

(ii) Parachutist A would have a higher terminal velocity *(1 mark)* because he has a greater weight. *(1 mark)* This means the force of air resistance would need to be greater to balance his weight, and air resistance is greater at higher speeds. *(1 mark)*

(c) Because the parachute increases their air resistance/drag. *(1 mark)*

2 (a) The upwards force must be greater because Stefan is accelerating upwards. *(1 mark)*

(b) Assume g = 10 N/kg, Stefan's mass is 600 ÷ 10 = 60 kg *(1 mark)* Force = mass × acceleration = 60 × 2.5 = 150 N *(1 mark)*

3 (a) Newton's third law says that if an object A exerts a force on another object, B, then object B exerts an equal but opposite force on object A. *(1 mark)* So if the bat exerts a force of 500 N on the ball, the ball also exerts a force of 500 N on the bat, but in the opposite direction. *(1 mark)*

(b) The ball's acceleration is greater *(1 mark)* because it has a smaller mass than the bat (F = ma). *(1 mark)*

Page 120

Warm-Up Questions

1) The distance travelled in the time between a hazard appearing and the driver braking.

2) Braking distance

3) Momentum = mass × velocity

4) Any two of, e.g. seat belts / air bags / crumple zones.

5) Any one of, e.g. traction control / anti-lock brakes (ABS).

Exam Questions

1 (a) (i) Accept answers between 12 and 13 m *(1 mark)*

    (ii) 35 m *(1 mark)*

    (iii) 35 m – 12 m = 22 m *(1 mark)*

  (b) Braking distance *(1 mark)*
    Using the graph, thinking distance is about 15 m and braking distance about 38 m.

  (c) No, *(1 mark)* if stopping distance and speed were proportional the relationship between them would be shown by a straight line. *(1 mark)*

2 (a) (i) Momentum = mass × velocity = 100 × 6 = 600 kg m/s to the right *(1 mark)*

    (ii) 80 × 9 = 720 kg m/s to the left *(1 mark)*

  (b) (i) Take left as positive, then the momentum of the two players is 720 – 600 = 120 kg m/s. *(1 mark)* The mass of the two players is 100 + 80 = 180 kg, so the speed is 120 ÷ 180 = 0.67 m/s *(1 mark)*

    (ii) Left *(1 mark)*
    The two players travel in the direction player B was going because they had more momentum before the collision.

3 (a) To increase the time it takes for the person to stop moving and absorb some of their kinetic energy. *(1 mark)*

  (b) Active safety features interact with the car driver to help avoid a crash. *(1 mark)* Passive safety features do not interact with the driving of the car, but help to keep the people in the car safe if it crashes. *(1 mark)*

Page 126

Warm-Up Questions

1) Because work done is a measure of energy transfer.

2) The energy an object has due to its height above the ground.

3) G.P.E = mass × gravitational field strength × height

4) It increases by a factor of 3^2 or 9.

5) At the lowest point all the roller coaster's potential energy has been converted to kinetic energy.

Exam Questions

1 (a) 40 000 kg × 1.05 m/s^2 = 42 000 N
    (2 marks, allow 1 mark for correct working)

  (b) 42 000 N × 700 m = 29 400 000 J *(1 mark)*

  (c) 29 400 000 J ÷ 29 400 N = 1000 m (1 km)
    (2 marks, allow 1 mark for correct working)

  (d) The train's kinetic energy would be reduced by a factor of 0.5^2 = 0.25, as would its braking distance, so the train would stop in 250 m. *(1 mark)*

  (e) Heat/sound energy *(1 mark)*

2 (a) 45 kg × 10 m/s^2 × 12 m = 5400 J
    (2 marks, allow 1 mark for correct working)

  (b) 5400 J *(1 mark)*
    The work done against gravity equals the change in potential energy.

3 (a) 529 000 J ÷ (10 m/s^2 × 26.45 m) = 2000 kg
    (2 marks, allow 1 mark for correct working)

  (b) (i) 529 000 J *(1 mark)*
    All the train's potential energy is converted to kinetic energy.

    (ii) v^2 = 2 × kinetic energy ÷ mass = 2 × 529 000 ÷ 2000 = 529, so v = 23 m/s *(2 marks, allow 1 mark for correct working)*

Page 130

Warm-Up Questions

1) Power

2) Accelerating needs a resultant force, so uses more fuel (to produce a greater driving force) than is needed to drive at a steady speed.

3) Because petrol and diesel will run out one day / damage the environment.

4) E.g. miles per gallon / litres per 100 kilometres.

5) By using electricity generated from fossil fuels.

Exam Questions

1 (a) Work done = mass × g × distance = 15 × 10 × 2 = 300 J
    (2 marks, allow 1 mark for correct working)

  (b) 300 J ÷ 12 s = 25 W *(2 marks, allow 1 mark for correct working)*

  (c) Electrical energy to kinetic energy *(1 mark)* to gravitational potential energy. *(1 mark)*

2 (a) Power = [0.5 × mass × velocity^2] ÷ time = [0.5 × 52 × 7^2] ÷ 3.5 = 1274 ÷ 3.5 = 364 W *(2 marks, allow 1 mark for correct working)*

  (b) Power = [mass × g × height] ÷ time = [52 × 10 × 0.2] ÷ 0.7 = 104 J ÷ 0.7 s = 149 W
    (2 marks, allow 1 mark for correct working)

  (c) [52 × 10 × 7] ÷ 45 = 3640 ÷ 45 = 81 W
    (2 marks, allow 1 mark for correct working)

3 (a) Fuel consumption = amount used ÷ distance travelled = 15.5 ÷ (240 ÷ 100) = 6.46 litres per 100 km *(1 mark)*

  (b) E.g. by increasing the weight of the car / by increasing the air resistance acting on the car. *(1 mark for each correct answer)*

  (c) Power = energy transferred ÷ time taken = 4800 ÷ (1 × 60) = 80 kW
    (2 marks, allow 1 mark for correct working)

Page 131

Revision Summary for Section Four

2) a) Speed = distance ÷ time = 3.2 ÷ 35 = 0.091 m/s
   b) Distance = speed × time = 0.091 × 25 × 60 = 137 m

3) No, the car was travelling at 12.6 m/s, so it wasn't speeding.

5) b) Acceleration = change in velocity ÷ time = (14 – 0) ÷ 0.4 = 35 m/s^2

13) b) Acceleration = force ÷ mass = 30 ÷ 4 = 7.5 m/s^2

20) Momentum = mass × velocity = 78 × 23 = 1794 kgm/s

23) Work = force × distance = 535 × 12 = 6420 J

24) PE = mass × g × height = 12 × 10 × 4.5 = 540 J

26) 150 kJ

27) KE = 0.5 × mass × velocity^2 = 0.5 × 600 × 40^2 = 480 000 J
    Height = PE ÷ (mass × g) = 480 000 ÷ (600 × 10) = 80 m

29) Power = energy transfer ÷ time = 540 000 ÷ 4.5 × 60 = 2000 W

30) Power = [mass × g × height] ÷ time = [78 × 10 × 20] ÷ 16.5 = 945 W

Pages 138-139

Warm-Up Questions

1) Positive and negative

2) Repel

3) Connecting an object to the ground using a conductor.

4) Any one of, e.g. in a thunder storm / in a grain chute / between paper rollers / when refuelling a vehicle.

5) Electrons.

Exam Questions

1 (a) A *(1 mark)* The rod is negatively charged so would repel the negative charges in the balloon, making them move away from the rod. *(1 mark)*

  (b) Because they are fixed in place so cannot move. *(1 mark)*

  (c) The negative charges in the rod attract the positive charges in the balloon. *(1 mark)* As Jane brings the rod closer to the balloon this attraction gets stronger, causing the balloon to move. *(1 mark)*

  (d) (i) The negative charges flow through Jane to Earth. *(1 mark)*
    (ii) Positively charged *(1 mark)*

2 (a) Electrons are scraped from the cloth onto the surface. *(1 mark)*

  (b) The charged cloth attracts tiny neutral dust particles. *(1 mark)*

3 (a) (i) Jyoti might get an electric shock from the chair leg. *(1 mark)*

    (ii) Jyoti's shoes insulate her from Earth so she will become statically charged when she walks on the carpet. *(1 mark)* This charge will flow to earth when she touches the conducting material. *(1 mark)*

  (b) (i) Touching the metal chair leg will connect Jyoti to Earth and allow any charge that has built up on her to disperse. *(1 mark)*

    (ii) No, *(1 mark)* because wood is an insulating material so would not allow the charge to flow to Earth. *(1 mark)*

Rubbing the bristles against the cloth will cause a build up of static electricity, *(1 mark)* which will attract the gold leaf. *(1 mark)*

(a) Restarting people's hearts. *(1 mark)*

(b) Insulating, so that the only person to be given a shock is the patient. *(1 mark)*

(c) Any one of, e.g. paint sprayer / dust precipitator / photocopier. *(1 mark)*

(a) Raindrops and ice particles bumping against one another. *(1 mark)*

(b) The top of the cloud becomes positively charged and the bottom becomes negatively charged. *(1 mark)*

(c) Lightning rods conduct the current from a charged cloud directly to Earth. *(1 mark)* The very high current doesn't pass through the fabric of the building, so there's less risk of fires starting. *(1 mark)*

Page 146

Warm-Up Questions

) Ohms, Ω

) It decreases.

)

) 50 Hz

) $1 \div 100 = 0.01$ s

Exam Questions

(a) (i) and (ii)

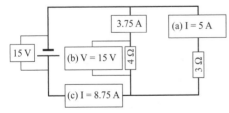

(1 mark each for the three components correctly drawn and placed)

(1 mark for correctly drawn arrows)

(b) Voltage = current × resistance = 0.3 × 5 = 1.5 V *(1 mark)*

(a) Diodes only allow current to flow in one direction. *(1 mark)*

(b) Resistance = voltage ÷ current = 6 ÷ 3 = 2 Ω
(2 marks, allow 1 mark for correct working)
Read the values from the graph, then use the formula R = V ÷ I.

(a) The trace shows an AC source so cannot be from a battery / must be from mains electricity. *(1 mark)*

(b) 20 ms *(1 mark)*
The wave takes four divisions to repeat. 4 × 5 ms = 20 ms.

(c) 20 ms = 0.02 s. Frequency = 1 ÷ time = 1 ÷ 0.02 = 50 Hz *(1 mark)*

(d) The amplitude of the wave will be decreased so the peaks and troughs will be smaller, *(1 mark)*

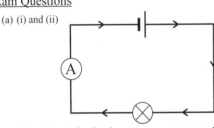

(a) Current = voltage ÷ resistance = 15 ÷ 3 = 5 A
(2 marks, allow 1 mark for correct working)

(b) 15 V *(1 mark)*
Voltage is the same across each branch in a parallel circuit.

(c) 5 + 3.75 = 8.75 A *(2 marks, allow 1 mark for correct working)*

Page 152

Warm-Up Questions

) The neutral wire (also accept the earth wire).

) Electrical energy (to kinetic energy) to heat energy.

) Current

4) E (energy) = Q (charge) × V (voltage).

5) The earth wire.

Exam Questions

1 (a) (i) brown *(1 mark)*

(ii) blue *(1 mark)*

(iii) green and yellow stripes *(1 mark)*

(b) The live and neutral wires. *(1 mark)*

(c) A fuse. *(1 mark)*

2 (a) (i) Current flows in and out through the live wire and neutral wires. No current flows in the earth wire. *(1 mark)*

(ii) A large current flows in through the live wire and out through the earth wire. *(1 mark)*

(iii) No current flows until the fuse is replaced. *(1 mark)*

(b) It would flow from the live wire through the person to Earth. *(1 mark)*

(c) A residual current circuit breaker cuts off the power *(1 mark)* when it detects that the current in the live wire is bigger than the current in the neutral wire. *(1 mark)*

3 (a) Power = current × voltage = 0.5 × 3 = 1.5 W
(2 marks, allow 1 mark for correct working)

(b) Charge = current × time = 0.5 × (30 × 60) = 900 C
(2 marks, allow 1 mark for correct working)

(c) Energy = charge × voltage = 900 × 3 = 2700 J
(2 marks, allow 1 mark for correct working)

Page 153

Revision Summary for Section Five

7) The wave should be drawn with one full wavelength (peak to peak) covering ten divisions of the screen.

8) $A_1 = 0.2$ A, $V_1 = 1.4$ V, $V_2 = 0.4$ V, $V_3 = 10.2$ V, $R_1 = 51 \Omega$

10) $A_1 = 2.4$ A, $A_2 = 1.2$ A, $V_1 = V_2 = 12$ V

14)a) Current = power ÷ voltage = 1100 ÷ 230 = 4.7 A
A 5 A fuse should be used.

b) 2000 ÷ 230 = 8.7 A
A 13 A fuse should be used.

15) Charge = current × time = 3 × (60 + 20) × 60 = 14 400 C

16) Energy = charge × voltage = 530 × 6 = 3180 J

Page 160

Warm-Up Questions

1) Vector quantities have magnitude (size) and direction, whereas scalar quantities only have magnitude.

2) Any two vector quantities, e.g. displacement/velocity/acceleration/force/momentum.
Any two scalar quantities, e.g. mass/time/temperature/length/speed/distance.

3) By finding the difference between the speeds of the two objects.

4) $s = \dfrac{(u + v)}{2} t$ and $s = ut + \dfrac{1}{2}at^2$

5) Gravity only.

Exam Questions

1 (a) Average speed = total distance ÷ total time
= (192 + 3600) ÷ [(2 + 3) × 60] = 3792 ÷ 300 = 12.64 m/s
(2 marks, allow 1 mark for correct working)

(b) (i) 0.013 − 0.011 = 0.002 m/s *(1 mark)*

(ii) 0.013 + 0.011 = 0.024 m/s *(1 mark)*

2 (a)

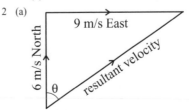

(1 mark for each vector correctly drawn)

(b) Speed: $\sqrt{6^2 + 9^2}$ *(1 mark)* $= \sqrt{117} = 10.8$ m/s (to 1 d.p.) *(1 mark)*
Direction: tan θ = 9 ÷ 6, so θ = tan⁻¹(1.5) *(1 mark)*
= 56° (to nearest degree) *(1 mark)*

3 (a) (i) v = u + at = 9 + 7 × 3 = 30 m/s
(2 marks, allow 1 mark for correct working)

(ii) $s = \dfrac{(u + v)}{2}t = \dfrac{(9 + 30)}{2} \times 3 = 58.5$ m *(1 mark)*

(b) (i) s = ut + 0.5at² = 0 × 1.73 + 0.5 × 10 × 1.73² = 0 + 14.96
= 14.96 m (to 2 d.p.) *(2 marks, allow 1 mark for correct working)*

(ii) v = s ÷ t = 2.59 ÷ 1.73 = 1.50 m/s *(1 mark)*

Page 165

Warm-Up Questions

1) Multiply the force by the perpendicular distance from the pivot.

2) Nm (newton metres)

3) There is a resultant moment so it turns.

4) The point where the object's whole mass can be considered to be 'concentrated'.

5) Because there is a resultant moment (caused by the weight acting out of line with the pivot).

Exam Questions

1 (a) (i) Moment = force × distance from pivot = 15 × 0.03 = 0.45 Nm
(2 marks, allow 1 mark for correct working)

(ii) 15 × 0.12 = 1.8 Nm *(2 marks, allow 1 mark for correct working)*

(b) End B should be put into the bolt *(1 mark)* because the same force exerts a larger moment (because it allows a larger distance between the pivot and the point where the force is applied). *(1 mark)*

2 (a) The decoration will hang with its centre of mass directly below the point of suspension. *(1 mark)*

(b) Maurice should suspend the decoration and a plumb line from the same point. *(1 mark)* When they stop moving, he should draw a line on the decoration where the plumb line lies. *(1 mark)* He should then repeat this, but with the shape suspended from a different pivot point. *(1 mark)* The centre of mass is where the two lines cross. *(1 mark)*

3 (a) No Robert is not correct. *(1 mark)* A seesaw will balance when the moments acting on each side of the pivot are equal (the masses only have to be equal when they are at the same distance from the pivot). *(1 mark)*

(b) (i)

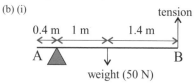

(1 mark for a correctly placed pivot)
(1 mark for each correctly placed and labelled force arrow)

(ii) Clockwise moments = anticlockwise moments.
50 × 1 = T × (1.4 + 1). T = 20.8 N (to 1 d.p.)
(2 marks, allow 1 mark for correct working)

Page 170

Warm-Up Questions

1) Centripetal force

2) Gravity

3) Any two of, e.g. communication (telephone, television) / monitoring weather or climate / spying / navigation.

4) Force due to gravity = 1 / distance²

5) When it is furthest from the Sun.

Exam Questions

1 (a) Because its direction (and therefore velocity) is constantly changing.
(1 mark)

(b) (i) Any two of, e.g. driving force / friction / weight / air resistance.
(1 mark)

(ii) The forces must be unbalanced *(1 mark)* because the car is accelerating. *(1 mark)*
Anything accelerating must have a resultant force acting on it.

---

2 (a) Geostationary/geosynchronous *(1 mark)*

(b) 24 hours *(1 mark)*

(c) (i) Satellite B sweeps over both poles while the Earth turns beneath it. *(1 mark)* So on each orbit the satellite 'sees' a different area of the Earth. *(1 mark)*

(ii) B — it must be in a low polar orbit (so that it can take quite detailed photographs of weather systems). *(1 mark)*

3 (a) Venus travels faster *(1 mark)* because it has a larger centripetal force acting on it. *(1 mark)*

(b) 1.39² = 1.9321 *(1 mark)* So the force of the Sun's gravity is 1.93 times weaker on Earth than it is on Venus. *(1 mark)*

Page 171

Revision Summary for Section Six

2) Average speed = total distance ÷ total time
= [2 + 10] ÷ [(15 + 30) ÷ 60] = 12 ÷ 0.75 = 16 mph

3) 0.5 − 0.2 = 0.3 m/s south

4) v² = 12² + 5² = 169. v = 13 mph
tan⁻¹(5 ÷ 12) = 22.6°
The bird is flying 13 mph at 22.6° west of north.

5) $s = \dfrac{(u + v)}{2}t = [(0 + 14) ÷ 2] \times 0.4 = 2.8$ m

7) a) s = ut + $\dfrac{1}{2}$at²   = 0 × 10 + 0.5 × 10 × 10² = 500 m

b) s = v × t = 1.5 × 10 = 15 m

10) 600 × 1.5 = 450 × d. d = 900 ÷ 450 = 2 m

Page 178

Warm-Up Questions

1) The direction of the magnetic field around a current carrying wire.

2)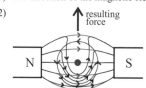

3) First finger — field
Second finger — current
Thumb — motion

4) Any two of, e.g. electric motors / speakers / generators / food mixers / CD players / locks.

5) Number of turns on coil, strength of magnetic field, amount of current, presence of magnetic/iron core.

Exam Questions

1 (a)

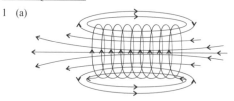

(1 mark for correct field shape, 1 mark for direction)

(b) E.g. soft iron *(1 mark)*

(c) A magnetically soft material loses its magnetism when the current is switched off. *(1 mark)*

(d) It will increase the strength of the magnetic field produced by Arnold's electromagnet. *(1 mark)*

2 (a) Anticlockwise *(1 mark)*
Use Fleming's left-hand rule on one of the arms of the coil.

(b) The split-ring commutator keeps the motor turning in the same direction. *(1 mark)*

(c) Any one of, increase the number of turns on the coil / increase the current flowing through the coil / use a stronger magnet / add an iron core to the coil. *(1 mark)*

(d) By reversing the direction of the current / the magnetic field. *(1 mark)*

Page 183

Warm-Up Questions

Slip-rings and brushes.

$$\frac{V_p}{V_s} = \frac{N_p}{N_s} \text{ or } \frac{V_s}{V_p} = \frac{N_s}{N_p}$$

Isolating transformers are transformers which have equal numbers of turns on their primary and secondary coils. They don't change the voltage of the supply — they're only used for safety.

To transmit a lot of power efficiently (by keeping the current low).

Exam Questions

A coil of wire is rotated inside a magnetic field. / A magnet is rotated within a coil of wire *(1 mark)* As the coil/magnet spins a voltage is induced within the coil. *(1 mark)* This voltage and the resulting current reverse direction with every half-turn of the coil/magnet — resulting in an alternating current. *(1 mark)*

(a) (i) B *(1 mark)*
   (ii) A *(1 mark)*
     Turning the magnet faster increases the peak voltage as well as frequency.

(b) (i) No voltage is induced. *(1 mark)*
   (ii) The lights would go out when the rider stops moving, e.g. at a junction. *(1 mark)*

(a) A voltage will not be induced in the secondary coil when using a DC supply *(1 mark)* because the magnetic field generated in the iron core is not changing. *(1 mark)*

(b) (i) Power = current × voltage = 2.5 × 12 = 30 W *(1 mark)*
   (ii) Assuming that the transformer is 100% efficient, the output power is 30 W. *(1 mark)* Current = power ÷ voltage = 30 ÷ 4 = 7.5 A *(1 mark)*
   (iii) $\frac{V_s}{V_p} = \frac{N_s}{N_p} = \frac{4}{12} = \frac{N_s}{15}$. $N_s = 60 ÷ 12 = 5$ turns
     (2 marks, allow 1 mark for correct working)

Page 187

Warm-Up Questions

Nuclear energy to heat energy to kinetic energy to electrical energy.

E.g. uranium and plutonium.

It produces a lot of radioactive waste that must be carefully disposed of.

It would have made it possible to generate a lot of electricity easily and cheaply.

The results have not been repeated reliably enough.

Exam Questions

(a) Any two of, e.g. risk of radioactive leaks / risk of major catastrophe (meltdown) / dangers of transporting fuel and waste to and from the plant / increased traffic to and from the plant. *(1 mark each)*

(b) Any two of, e.g. bring skilled job opportunities to the area / bring money into the area / a reliable supply of electricity. *(1 mark each)*

(c) Any one of, e.g. Nuclear power is more environmentally friendly in that it does not release emissions that add to the greenhouse effect or cause acid rain. / Nuclear power is less environmentally friendly in that it creates radioactive waste which is hard to dispose of safely / leaks radioactive materials into the environment. *(1 mark)*

(d) (i) A slow-moving neutron is absorbed by a uranium or plutonium nucleus, causing the nucleus to split. *(1 mark)* When a nucleus splits up, it releases two or three neutrons. *(1 mark)* These might go on to hit other nuclei, causing them to split and release even more neutrons, starting the process over again. *(1 mark)*
   (ii) Control rods are used in a nuclear power station *(1 mark)* to absorb excess neutrons and control the rate of the chain reaction. *(1 mark)*

(a) Deuterium and hydrogen *(1 mark)*
Fission uses heavy elements, whereas nuclear fusion uses light elements.

(b) Fusion power would allow a lot of electricity to be generated easily and cheaply *(1 mark)* without the large amounts of waste currently produced by fission. *(1 mark)*

(c) Fusion only works at such high temperatures that it uses more energy than it can produce. *(1 mark)*

Page 190

Revision Summary for Section Seven

13) Assume the transformer is 100% efficient.
   6 × 10 = 3 × I. I = 60 ÷ 3 = 20 A

14) $\frac{V_p}{V_s} = \frac{N_p}{N_s}$. $N_p = (20 × 64) ÷ 16 = 80$ turns

Page 195

Warm-Up Questions

1) A real image can be captured on a screen. A virtual one can't.

2) Angle of incidence = angle of reflection

3)
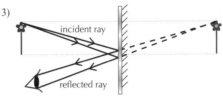

4)
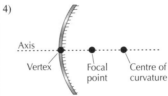

5) Whether the image is real or virtual, which way up it is (the same way up as the object or inverted) and what size it is (bigger, smaller or the same size as the object).

Exam Questions

1 (a) convex *(1 mark)*
   Convex mirrors make objects look smaller.

   (b) (i) 2 × 3 = 6 cm *(1 mark)*

     (ii)
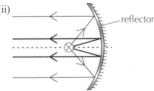

     (1 mark for two or more rays drawn correctly from bulb to reflector, and clearly showing parallel reflected rays. 1 mark for arrows on rays from bulb to reflector. 1 mark for arrows on reflected rays.)

   (c) The paintwork is a smooth/even surface, so parallel light rays all reflect at the same angle and Jennie's reflection is clear. *(1 mark)*

   The bumper is a rough/uneven surface, so parallel light rays reflect at different angles, giving a diffuse reflection. *(1 mark)*

2 As light reflected from the mobile phone leaves the pond it speeds up, because air is less dense than water *(1 mark)*. This causes it to bend away from the normal, making the mobile phone (and therefore the bottom of the pond) appear closer than it actually is, i.e. shallower *(1 mark)*. The phenomenon is called refraction *(1 mark)*.

3
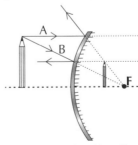

   (1 mark for ray A and its reflected ray. 1 mark for ray B and its reflected ray. 1 mark for rays extended backwards (dotted lines). 1 mark for image in correct position.)

Page 199

Warm-Up Questions

1) $n = \dfrac{\sin i}{\sin r}$

2) $n = \dfrac{\text{speed of light in a vacuum}}{\text{speed of light in the material}}$ or $n = \dfrac{c}{v}$

3) e.g.

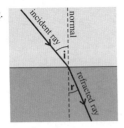

4) violet

5) $\sin C = \dfrac{n_r}{n_i}$

Exam Questions

1 (a) $n = \sin 54° \div \sin 33° = 1.49$ (to 2 d.p.) *(1 mark)*

(b) $\sin i = 1.56 \times \sin 25° = 0.659...$ *(1 mark)*
$i = \sin^{-1}(0.659) = 41°$ (to nearest degree) *(1 mark)*

(c) n = speed of light in vacuum ÷ speed of light in material
$= (3 \times 10^8) \div (1.89 \times 10^8) = 1.59$ (to 2 d.p.) *(1 mark)*

2 (a) $\sin C = 1 \div n_{water}$ or $n_{water} = 1 \div \sin C$ or $\sin C = 1 \div 1.33$
$= 0.752$
$C = \sin^{-1}(0.752)$
$= 49°$

(3 marks for correct answer — otherwise up to 2 marks for working)

(b) total internal reflection *(1 mark)*

(c) Any one of, e.g. optical fibres, communication (using optical fibres), endoscopes, binoculars, periscopes. *(1 mark)*

Page 205

Warm-Up Questions

1) Concave/diverging and convex/converging.

2) Convex lenses bulge outwards, concave lenses curve inwards.

3) Concave/diverging lenses always create virtual images.
Convex/converging lenses can create real or virtual images (depending on the position of the object).

4) The point where the rays that hit the lens parallel to the axis appear to come from once they have passed through the lens.

5) Magnification = image height ÷ object height

6) Any two of, e.g. magnifying glass, camera lens, spectacles for long-sightedness, projector.

Exam Questions

1 (a) E.g. dispersion means the separation of white light/light from the Sun into its component colours/frequencies/wavelengths *(1 mark)*. Red light slows down the least / violet light slows down the most in glass *(1 mark)*, so red light is refracted the least / violet light is refracted the most *(1 mark)*.

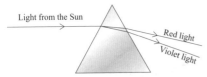

(1 mark for rays drawn correctly, showing red light refracted least and violet refracted most.)

(b) The boundaries where light enters and leaves the block are parallel, so the colours recombine when they leave the block. *(1 mark)*

2 (a) A concave lens causes parallel rays of light to diverge (spread out) rather than converge (come together) *(1 mark)*. This means that Edward's lens cannot focus the sunlight to start a fire *(1 mark)*.

(b) (i)

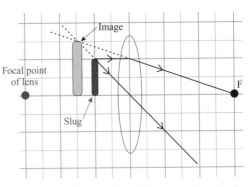

Ray going to correctly positioned focal point *(1 mark)*. Ray going through centre of lens *(1 mark)*. Both rays extended backwards (dotted) and image drawn where they cross *(1 mark)*.
You sometimes need to draw another focal point on the opposite side of lens, the same distance away from the centre line of the lens.

(ii) Magnification = image height ÷ object height = 3 ÷ 2 = 1.5
(1 mark for answer between 1.4 and 1.6)
Draw diagrams like this as neatly as you can so that you can measure the image and the object accurately.

Page 209

Warm-Up Questions

1) Constructive (displacements in the same direction) and destructive (displacements in opposite directions).

2) Add the amplitudes at each point taking into account the directions (add if the same direction, subtract if opposite directions).

3) one wavelength

4) diffraction

5) The 'filtering' of waves so that they're all vibrating in the same direction.

6) transverse waves

Exam Questions

1 (a) She might hear the sound getting louder and quieter several times as she walks along. *(1 mark)*

(b) interference *(1 mark)*

2 Light reflected from roads (especially wet roads) is mainly horizontally polarised *(1 mark)*. Vertically polarising glass filters out the horizontally polarised light to reduce the glare *(1 mark)*.

3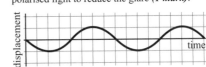

(1 mark for wave shown with correct phase (i.e. goes negative first).
1 mark for wave shown with correct frequency (i.e. same as the two waves given). 1 mark for correct amplitude (i.e. one square each side of x-axis).)

Page 217

Warm-Up Questions

1) longitudinal

2) 20 Hz to 20 000 Hz

3) a copper bar

4) higher frequency = higher pitch
larger amplitude = louder (higher volume)

5) distance = velocity × time

Exam Questions

1 (a) Distance = 2 × 360 = 720 m
$D = v \times t$, so $v = d \div t = 720 \div 0.48 = 1500$ m/s
(2 marks for correct answer — otherwise 1 mark for correct formula)
Don't forget to double the distance in echo calculations — the sound has to travel down to the sea bed and back again.

(b) 180 ms = 0.18 s

Distance = v × t = 1500 × 0.18 = 270 m

So the fish are 270 ÷ 2 = 135 m below the ship. *(2 marks for correct answer (based on the speed calculated in part (a)) — otherwise 1 mark for correct formula)*

(c) 50 kHz *(1 mark)*

(d) Any one of, e.g. prenatal foetal scanning, cleaning delicate items/teeth/electronics, industrial quality control/crack detection. *(1 mark)*

(a) e.g.

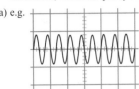

(1 mark for a sine wave with higher frequency)

(b) e.g.

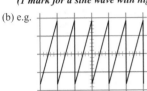

(1 mark for greater amplitude. 1 mark for the same frequency and shape)

Page 218

Revision Summary for Section Eight

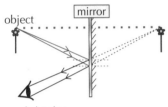

object   mirror

n = sin i ÷ sin r

= sin 30 ÷ sin 20

= 1.46 (to 2 d.p.)

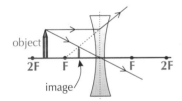

object

image

2F   F   F   2F

4) 4.5 ÷ 1.5 = 3

5) Distance travelled by pulse = 2800 × 0.003 = 8.4 m.

So depth of pipe = 8.4 ÷ 2 = 4.2 m.

Page 225

Warm-Up Questions

1) $V_{out} = V_{in} \times \left(\dfrac{R_2}{R_1 + R_2} \right)$

2) symbol: 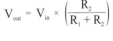   graph:

3) p-type and n-type

4) four

5) a capacitor

6) False — it stores charge.

Exam Questions

1 (a) $V_{out} = V_{in} \times \dfrac{R_2}{R_1 + R_2}$

= 12 × (210 ÷ (210 + 630))

= 12 × 210 ÷ 840 = 3 V

(2 marks for correct answer — otherwise 1 mark for correct working)

(b) Any one of, replace R₁ with a resistor of a lower value / replace R₂ with a higher value resistor. *(1 mark)*

(c) (i) approximately 12 V *(1 mark)*

(ii) approximately 0 V *(1 mark)*

(d) No current flows *(1 mark)* because the diode does not allow current to flow when connected that way round *(1 mark)*.

2 E.g. When the switch is closed, current flows *(1 mark)*, charge builds up on the capacitor, and the p.d. (voltage) across it increases *(1 mark)*. When the p.d. (voltage) becomes equal to that of the battery, current stops flowing *(1 mark)*. The capacitor is fully charged *(1 mark)*.

Page 231

Warm-Up Questions

1)

AND gate			OR gate		
Input A	Input B	Output	Input A	Input B	Output
0	0	0	0	0	0
0	1	0	0	1	1
1	0	0	1	0	1
1	1	1	1	1	1

NAND gate			NOR gate		
Input A	Input B	Output	Input A	Input B	Output
0	0	1	0	0	1
0	1	1	0	1	0
1	0	1	1	0	0
1	1	0	1	1	0

2) E.g. it's used to store a signal until it is reset / it's used as a memory. Latches use NOR gates.

3) Light emitting diode.

4) To isolate parts of a circuit with different voltages/power supplies e.g. high voltage and low voltage or A.C. and D.C.

Exam Questions

1 (a) Any one of, e.g. it uses less power / it lasts longer / it takes less current to run. *(1 mark)*

(b) Connect a resistor in series with it. *(1 mark)*

2 (a) (i) ON or 1 *(1 mark)*

(ii) E.g. when the low current/control circuit gives an output of 1, current flows through the coils of the electromagnet in the relay, making it magnetic *(1 mark)*.

An iron contact is attracted by the electromagnet *(1 mark)*.

A rocker pivots and closes the electrical contacts which allow current to flow in the high current circuit *(1 mark)*.

(b)

temperature	pump fault	main switch	A	output
1	1	1	0	0
1	1	0	0	0
1	0	1	0	0
1	0	0	0	0
0	1	1	0	0
0	1	0	0	0
0	0	1	1	1
0	0	0	1	0

(1 mark for each correct row — 7 marks maximum)

<u>Page 232</u>

<u>Revision Summary for Section Nine</u>

2   a) $30 \div (30 + 10) = 0.75$
          $0.75 \times 9 = \textbf{6.75 V}$

     b) $2 \div (2 + 10) = 0.167$
          $0.167 \times 9 = \textbf{1.5 V}$

9   a)   off

     b)   on

     c)   off

<u>Page 237</u>

<u>Warm-Up Questions</u>

1) $P \div T = $ constant   or $P_1 \div T_1 = P_2 \div T_2$

2) $PV \div T = $ constant   or $P_1 V_1 \div T_1 = P_2 V_2 \div T_2$

3) The particles are moving around and colliding with the walls.

4) 302 K, 243 K, 447 K, 273 K

5) –73 °C, 77 °C, –273 °C, 350 °C

<u>Exam Questions</u>

1   (a) Any two of, e.g. the aerosol will absorb heat energy from the sun / the kinetic energy of the gas molecules will increase / the gas molecules will move faster / the gas molecules will hit the walls of the can more often / kinetic energy is proportional to temperature / the temperature of the gas will increase / $P \div T$ is constant. *(1 mark each — maximum 2 marks)*

      The pressure of the gas will increase and the aerosol could explode. *(1 mark)*

   (b) No, it is not true. *(1 mark)*
At 0 K (absolute zero) the particles have the least possible amount of kinetic energy, so cannot be cooled any further. *(1 mark)*

   (c) E.g. there are less air molecules inside the bottle to collide with the inside so the pressure drops *(1 mark)*. The air molecules outside the bottle are colliding with it and exerting a pressure on the outside *(1 mark)*. The air molecules on the outside push the sides of the bottle in (until the pressures balance again) *(1 mark)*.

2   (a) Because when they are warm/hot the pressure will increase. *(1 mark)*

   (b) 15 + 273 = 288 K. *(1 mark)*

   (c) $P \div T = $ constant or $P_1 \div T_1 = P_2 \div T_2$

      $T_2 = P_2 T_1 \div P_1 = (35 \times 288) \div 31 = 325$ K.
      $T_2 = 325 - 273 = 52$ °C.
(3 marks for correct answer — otherwise 1 mark for equation, 1 mark for working)
P.s.i. is a unit of pressure. As long as you keep the units the same throughout the equation, it doesn't matter what units you use.

3   $P_1 V_1 \div T_1 = P_2 V_2 \div T_2$
so $V_2 = P_1 V_1 T_2 \div P_2 T_1$

    $V_2 = (2 \times 25 \times 333) \div (2.6 \times 293) = 21.9$ m³.
(3 marks for correct answer — otherwise 1 mark for equation, 1 mark for working)

<u>Page 242</u>

<u>Warm-Up Questions</u>

1) alpha (α) radiation — helium nuclei
beta-minus (β–) decay — electrons
beta-plus (β+) decay — positrons
gamma (γ) radiation — electromagnetic waves
neutron radiation — neutrons.

2) A positively charged particle with the same mass as an electron.

3) Because they are not charged, so are not directly ionising and can't be deflected by an electric field.

4) It has too many neutrons.
It has too few neutrons.
It is too heavy.
It has too much energy.

5) A particle with the same mass as another but an opposite charge.

6) Up-quark — relative charge = 2/3, relative mass = 1/3
Down-quark — relative charge = –1/3, relative mass = 1/3

7) Neutron — 1 up-quark and 2 down-quarks = 2/3 + (–1/3) + (–1/3) = 0
Proton — 2 up-quarks and 1 down-quark = 2/3 + 2/3 + (–1/3) = +1

<u>Exam Questions</u>

1   (a) (i) Any one of, water / concrete / polythene. *(1 mark)*
Neutrons are absorbed best by light nuclei (e.g. hydrogen). *(1 mark)*
The material suggested is a hydrogen-rich substance. *(1 mark)*

     (ii) Thick lead. Gamma rays are often emitted as well. *(1 mark)*

   (b) They can look for nuclear decays from nuclei that have been made radioactive by neutron radiation. *(1 mark)*

2   (a) 137 – 55 = 82 neutrons. Cs-137 therefore lies above the curve of stability *(1 mark)*. It therefore has too many neutrons *(1 mark)*, so undergoes beta-minus decay *(1 mark)*.

   (b) Alpha decay is most likely *(1 mark)*. U-238 has a heavy nucleus / proton number greater than 82 *(1 mark)*.

   (c) $^{22}_{11}\text{Na} \longrightarrow \ ^{22}_{10}\text{Ne} \ + \ ^{0}_{+1}\beta$

     (β+ particle — 1 mark. A second product with the correct numbers — mark. Correct identification of Ne-22 from table — 1 mark.)

   (d) An up-quark changes into a down-quark plus a positron *(1 mark)*, changing a proton into a neutron *(1 mark)*. The nucleus emits the positron *(1 mark)*.

<u>Page 246</u>

<u>Warm-Up Questions</u>

1) Thermionic emission

2) KE = charge on electron × accelerating voltage
or KE = e × V

3) Current (I) = charge passed (n × e) ÷ time (t)

4) 1.6×10^{-19} C

5) E.g. smaller charge on plates / faster-moving particles / smaller charge on particles / heavier particles.

6) Any three of, e.g. TV picture tubes / computer monitors (not LCD) / oscilloscopes / production of X-rays.

<u>Exam Questions</u>

1   (a) E.g. the cathode is heated, causing electrons to 'boil off' the cathode in thermionic emission *(1 mark)*. Electrons accelerate towards the anode, which allows them to pass through as a beam *(1 mark)*. Deflecting plates can change the path of the electrons *(1 mark)*. They finally hit the phosphorescent screen, creating a glow *(1 mark)*.

   (b) $I = (n \times e) \div t$
     $= (7.5 \times 10^{19} \times 1.6 \times 10^{-19}) \div 5$
     $= \textbf{2.4 A}$
(2 marks for correct answer — otherwise 1 mark for equation)

   (c) $KE = e \times V$
     $V = KE \div e$
     $= 3.2 \times 10^{-15} \div 1.6 \times 10^{-19}$
     $= \textbf{20 000 V}$ (or 20 kV)
(2 marks for correct answer — otherwise 1 mark for rearranging equation)

2   (a) (i) 6 *(1 mark)*

     (ii) 5 *(1 mark)*

   (b) 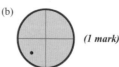 *(1 mark)*

   (c) The beam would move up and down *(1 mark)*, causing the screen to display a vertical line (if the frequency was great enough) *(1 mark)*.

<u>Page 247</u>

<u>Revision Summary for Section Ten</u>

3) a)   i) 184 K

      ii) 393 K

   b)   5 °C

2.90 atm (to 3 s.f.)

167 K (to 3 s.f.)

3.68×10^{-17} J (to 3 s.f.)

4 A

The electron beam will be deflected upwards and right (towards the positive plates).

ge 253

arm-Up Questions

Two bundles of optical fibres — one bundle to carry light into the body, one bundle to carry the image back to the eyepiece or TV screen.

pulse oximeter

Any three of, e.g. keeping warm, moving, digesting food, the working of organs, tissue repair.

food — it is chemical energy

EMG — electromyogram. Used to identify muscle problems / train stroke sufferers to use their muscles
ECG — electrocardiogram. Used to monitor heart rate / identify abnormalities in the heartbeat.

frequency (Hz) = 1 ÷ time period (s)

xam Questions

(a) Any one of, e.g. faster recovery time / less invasive / only needs a small hole in the body. *(1 mark)*

(b) an endoscope *(1 mark)*

(c) Any thin part of the body, e.g. the ear lobes or fingers. *(1 mark)*

(a) (i) It's the increased potential when an electrical signal passes along a muscle cell or nerve cell. *(1 mark)*

(ii) The muscle contracts. *(1 mark)*

(b) Time between peaks = 1.4 − 0.8 = 0.6 s
Heart rate = 1 ÷ 0.6 = 1.667 Hz
Rate per minute = 1.667 × 60 = 100 bpm
(2 marks for correct answer — otherwise 1 mark for working)

(c) The ventricles are relaxing. *(1 mark)*

(a) The rate the body transfers/burns energy at rest. *(1 mark)*

(b) Any one of, e.g. exercise to reduce fat and build up muscle / lose body fat slowly by dieting (fast weight loss reduces BMR). *(1 mark)*

age 258

/arm-Up Questions

Radiation is energy emitted from a source (as electromagnetic waves or moving particles).

Intensity = power ÷ area or I = P ÷ A

U-235 is bombarded with thermal/slow neutrons.

A circular particle accelerator used for making isotopes by proton enrichment (and also for particle physics experiments and radiation therapy).

Many of the isotopes used in hospitals have a short half-life so they have to be made close to the place they are used.

xam Questions

(a) Intensity = power ÷ area = 60 ÷ 3 = 3 = 20 W/m². *(1 mark)*

(b) Distance is doubled, so the same radiation is spread over four times the area (because area corresponds to distance²).
So, intensity = 20 ÷ 4 = **5 W/m²**
(2 marks for correct answer — otherwise 1 mark for quadrupling area)

(c) The air (along with any dust or other particles carried by it) will absorb some of the radiation. *(1 mark)*

(a) U-235 is bombarded with thermal/slow neutrons *(1 mark)*. A U-235 nucleus captures a neutron and changes into U-236 *(1 mark)*. The U-236 nucleus is unstable and splits into two smaller nuclei *(1 mark)*.

(b) $^{235}_{92}\text{U} + ^{1}_{0}\text{n} \longrightarrow ^{141}_{56}\text{Ba} + ^{92}_{36}\text{Kr} + 3^{1}_{0}\text{n}$ (+ energy)

(3 marks — 1 for the left-hand side, 1 for the two resulting nuclei, 1 for the three neutrons)

3 (a) positrons *(1 mark)*

(b) $^{18}_{8}\text{O} + ^{1}_{1}\text{p} \longrightarrow ^{18}_{9}\text{F} + ^{1}_{0}\text{n}$
(2 marks — 1 for left-hand side, 1 for right-hand side)

(c) Positron emitters are used in PET scans. *(1 mark)*

Page 264

Warm-Up Questions

1) momentum = mass × velocity

2) 1. Two things bouncing off each other.
2. Two things colliding and combining.
3. An explosion/shot and recoil/emission of a particle from a nucleus.

3) It means complete destruction and it happens when a particle meets its antiparticle (e.g. positron and electron) (the resulting energy is given off as radiation).

4) It must be a positron emitter with a short half-life (it must also be nontoxic).

5) E.g. a consensus is a collective opinion or general agreement in society. It is relevant because there can be moral and ethical questions about the medical treatments which science makes available.

Exam Questions

1 (a) Any two of, e.g. damaged heart tissue / coronary artery disease / previous heart attacks / cancer / tumours / blood flow problems in the brain / thyroid disfunction / Parkinson's disease / epilepsy / depression / Alzheimer's disease. *(1 mark each)*

(b) Any two of, e.g. destroy cells completely / damage cells / damage DNA/ genetic material / cause mutations / cause cells to become cancerous/ divide uncontrollably. *(1 mark each)*

(c) (i) C *(1 mark)* — it is a gamma emitter and has a reasonably long half-life *(1 mark)*.

(ii) E *(1 mark)* — it emits positrons and has a short half-life (but not too short like D — this would decay too quickly to travel around the body and be detected in the scanner) *(1 mark)*.

2 (a) zero before and after *(1 mark)*
The velocities at the beginning are equal but in opposite directions, so they cancel each other out. Momentum has to be conserved, so there's no momentum at the end either.

(b) All the mass of the particles is converted to energy *(1 mark)*. The kinetic energy changes to electromagnetic energy/gamma rays *(1 mark)*.

(c) annihilation *(1 mark)*

3 momentum before = momentum after *(1 mark)*
$(1 \times 14\,000) = (1 \times -13\,000) + (235 \times v_2)$ *(1 mark)*
$14\,000 = 235v_2 - 13\,000$
$v_2 = (14\,000 + 13\,000) \div 235$
$= 115$ km/s to the right *(1 mark) (3 marks in total)*

Page 265

Revision Summary for Section Eleven

10)a) 60 ÷ 0.7 = 85.71 bpm

12) I = P ÷ A = 1000 ÷ 4 = 250 W/m²

15) $^{235}_{92}\text{U} + ^{1}_{0}\text{n} \longrightarrow ^{145}_{57}\text{La} + ^{88}_{35}\text{Br} + 3^{1}_{0}\text{n}$ (+ energy)

16) positrons

17) The pink ball's initial momentum is (2 × 0.15 =) 0.3 kg m/s.
The yellow ball's initial momentum is (1 × 0.15 =) 0.15 kg m/s in the opposite direction. The total momentum after the collision is therefore 0.15 kg m/s. The yellow ball must therefore be moving at **1 m/s** in the original direction of the pink ball.

Index

A

absolute temperature scale 8
absolute zero 234
absorption 52
AC - alternating current 37, 41, 43, 142, 143, 179, 180, 221-224
AC generators 179
accelerating voltage 244
acceleration 102-105, 109, 110, 157, 166
acid rain 24, 129
active safety features 119
air bags 117, 118
air pollution 98
air resistance 112, 113, 124
alpha radiation 72, 73, 79, 81, 238, 240, 260
ammeters 38, 140, 145
amplitude 49, 51, 65, 206, 207, 211, 213
amps 39
analogue signals 65, 66
AND gates 226, 228, 229
angle of incidence 54, 190, 196, 197
angle of reflection 190
angle of refraction 196, 197
annihilation 260
anodes 243
anomalous results 267, 269
anticlockwise moments 163, 164
antimatter 241, 257
antiparticles 260
aquaplaning 115
artificial satellites 89
assumptions 84
asteroids 86
astronauts 96
atmosphere 74, 99
atmospheres (atm) 235
atomic bombs 256
atoms 238, 239
  structure of 71, 241
aurora borealis 89
average 5, 267, 268
average speed 154

B

background radiation 74, 75, 77, 89
balanced forces 109
balanced moments 163
bar magnets 88
basal metabolic rate (BMR) 250
bats 216
batteries 223
beam of electrons 243-245
bell jar experiment 211
beta radiation 72, 73, 79, 81, 238
beta-plus radiation 240
bias 5
Big Bang 93
big crunch 94
biogas 30
biomass 30
black holes 90, 92
blood 249
body fat 250
bombs 256
boundaries between media 214

brain scanning (PET) 261
brakes 115, 124
braking distances 115
bridge circuits 222
Bruce Willis 87
burglar detectors 39

C

calibration tables 84
cameras 204, 224
cancer 61, 79, 82, 261-263
capacitors 223, 224
car safety 118
carbon-11 257
carbon-14 83
carbon dating 84
cathode ray oscilloscope 37, 142
cathode ray tube 243
cavity wall insulation 17
CD players 63
cell damage 58
cells 61, 144
Celsius scale of temperature 234
central heating 9, 39
centre of curvature 192, 193
centre of mass 162
centripetal forces 166, 169
CFCs 58
charge 223
charged particles 244
chemical energy 19, 250
circuit breakers 147
circuits 219
circular motion 156
climate change 24, 129
clockwise moments 163, 164
coal 24
coil 173, 179
cold fusion 186
collisions 116-118, 259
combining vectors 155
comets 87, 169
communication 60
  satellites 167
concave mirrors 192, 193
condensing 10
conduction 13, 61
conductors 13, 147
conservation of momentum 116
constructive interference 206, 207
continuous data 268
controlled experiments 3, 4, 266
control rods 185
convection 14, 21, 61, 88
conventional current 37
converging lenses 201, 202, 204
convex lenses 204
convex mirrors 192, 194
cooking 57
cooling systems 9
core (of the Earth) 68
correlation 269
cosmic rays 74, 89, 93, 96
cost 47
coulombs 151
count rate 77
crash tests 119
critical angle 54, 197, 198, 248
CRO traces 179, 210, 212-214
crumple zones 117, 118

current 37-40, 43, 89, 140, 144, 145, 147, 149, 151, 176, 177, 221, 230, 244
curtains 17
curve of stability 239
curved mirrors 192, 193
cyclotrons 257

D

dam 33
dark matter 94
data 4, 267, 269
DC - direct current 37, 41, 142, 143, 221
decay, radioactive 71
decommissioning 26
defibrillators 137, 251
deflection 72, 244
  of electron beams 243-245
dependent variables 266
destructive interference 206, 207
detecting neutrons 239
diagnosis 263
diesel 129
diet 250
diffraction 52, 60, 208
diffuse reflections 190
digital signals 65, 66
digital systems 226
dimmer switches 65
diodes 141, 221, 222
discrete data 268
dispersion 197
displacement 157, 206
distance-time graphs 104
diverging lenses 201, 203
diverging light rays 189, 191
DNA 58
double glazing 17
double insulation 148
down-quark 241
drag 112, 113
draught-proofing 17
drugs 263, 266
dummy pills 266
dust particles 135
dust precipitators 136
dynamo effect 40, 41
dynamos 179

E

$E = mc^2$ 260
eardrum 210
Earth 88
  atmosphere of 60
earth wire 147, 148
earthing 135, 148
earthquakes 67, 68
  prediction of 68
Earth's structure 67
ECGs (electrocardiographs) 252, 263
echoes 215
economic factors 7
  in medical research 263
eddy currents 181
efficiency 20-22, 42, 43
elastic potential energy 19
electric circuits 149, 221
electric current 133, 172
electric fields 72, 244
electric shocks 148

electrical appliances 46, 47
electrical cables 147
electrical charge 151
electrical conductors 40
electrical disturbances 66
electrical energy 19, 40, 46, 150
electrical insulators 221
electrical oscillations 213
electrical power 47, 150
electrical pressure 37
electrical signals 63, 65
electricity 37, 40, 42, 43, 47, 221, 251, 252
electromagnet 173, 230
electromagnetic induction 40, 179, 180
electromagnetic radiation 57, 58, 93
electron beams 243-245
electron guns 243-245
electrons 37, 71, 72, 89, 132, 221, 241
electrostatic paint sprayers 136
EM waves 50, 65
EMG (electromyography) 251
endoscopes 248, 263
energy 8, 47, 49, 73, 79, 149, 151, 250, 254, 260
energy conservation 19
energy converted into mass 241
energy created from mass 260
energy in circuits 149
energy of waves 149, 212
energy resources 25
energy transfer 121, 128
energy transformation diagrams 151
environmental damage 25, 33
environmental factors 7, 24, 32, 185
equal and opposite force 111
equations of motion 157
escape lanes 118
ethics 6, 62, 263
evidence 1, 4, 5
exercise 250
expansion 93, 94
experiments 2, 3, 266-269
explosions 259
exposure, radiation 79
eyes 189, 190, 204

F

$F = ma$ 110
fair test 266
field diagrams 172
filament lamp 38
fission 256
Fleming's left-hand rule 175, 177
fluorescent tubes 58
fly-by missions 97
focal length 204
focal point 192-194, 201-203
foetus, prenatal scans 216
forces 156, 166, 174, 176
  centripetal 166, 169
  turning 161-164
fossil fuels 24, 42, 185
fossils 87
free-fall 113

Index

freezing 10
frequency 41, 49-51, 57, 65, 143,
frequency of sound waves 211-213
friction 112, 113, 166
fuel consumption 129
fuels for cars 129
full-wave rectification 222
fundamental particles 241
fuses 147, 148, 150
fusion 186

G

G-M tubes 77
gain 142
galaxies 90, 93
gamma radiation 50, 73, 79, 81, 82, 89, 238, 240, 260-262
gas-cooled nuclear reactor 185
gas-fired power stations 33
gases, kinetic theory of 233-236
generators 40, 41, 179
geostationary (geosynchronous) satellites 167
geothermal energy 25, 27
Global Positioning System (GPS) 167
global warming 185
gradient 104
grain chutes 134
graphs 269
gravitational field strength (g) 106, 122
gravitational potential energy 19, 122, 125
gravity 90, 92, 94, 106, 159, 169
greenhouse effect 24, 185, 228, 229

H

haemoglobin 249
half-life 76, 77, 81, 83, 257, 261, 263
half-wave rectification 222
Hawaii 98
hearing range for humans 211, 213
heart 251, 252, 261, 263
heat energy 10, 19, 149, 250
heat exchanger 185
heavy elements 92
helium nuclei 72, 92
hertz (Hz) 41, 49, 211
high voltage cables 133
highway code 115
horizontal motion 159
hospitals 257
Hubble 98, 168
hydroelectric power 32
hydrogen 92, 186
hypotheses 1, 2

I

ideal gases 236
images 189-191, 193, 194, 201-204
immersion heaters 14
incident rays 191, 193, 194, 196, 201, 203
independent variables 266

industrial quality control 215
infrared 15, 50, 57, 62, 255
input energy 20, 21
inputs 228, 229
insulation 17, 132
insulators 13, 147
intensity of radiation 254, 255
interference 51, 66, 206-208
inverse square relationship 169, 255
inverted images 189, 204
ionisation 72, 73, 79
ions 72, 89
iris scanning 63
iron core 181
isolating transformers 182
isotopes 71, 76, 257, 261

J

joules 8
junctions 145

K

Kelvin scale of temperature 234
keyhole surgery 248
kilowatt-hours 47
kinetic energy 13, 19, 40, 123-125, 233, 250
kinetic theory 233-236

L

lasers 63
latches 229
latent heat 11
law of reflection 190, 191
laws of motion 109-111
lenses 200-204
light 53, 63, 189, 190, 192, 196-198, 200, 201, 204, 207, 208
light-emitting diodes (LEDs) 230
light energy 19
light pollution 98
light years 90
light-dependent resistors 39
lightning 134
lines of best fit 269
live wire 147
loft insulation 17
logic circuits 230
logic gates 226-230
longitudinal waves 49, 67, 210
loudness 211
loudspeakers 63, 176
low polar orbits 168
lubricant 112
lung cancer 75

M

magnet 40
magnetic fields 40, 41, 72, 74, 88, 172, 174, 176, 179
magnetic poles 88
magnetically soft materials 173
magnification 189, 204
main sequence star 92
mains electricity 37, 142
manned spacecraft 96
mantle (of the Earth) 68
mass 94, 106
mass converted into energy 260

mass created from energy 241
matter and antimatter 241
mean 267
medical research 263
medical tracers 256
medical trials 263
medicine 261
melting 10, 11
metabolic rate 250, 261
meteorites 86
meteors 86
microwaves 50, 57, 61
Milky Way 90
mirrors 189, 191-194, 204
mobile phone communications 60
moments 161
momentum 116, 117
    conservation of 259, 260
moons 86
morals 6
motion 158, 177
motors 177
muscles 96, 250, 251
mutant cells, cancer 79
mutations 262

N

NAND gates 227
National Grid 42, 43, 182
natural gas 24
Near Earth Objects (NEOs) 87
negative charge 133
negative correlation 269
neutral wire 147
neutron rich isotopes 239
neutron star 92
neutrons 71, 72, 184, 239, 256
Newton, Isaac 109
night-vision equipment 62
nitrogen-13 257
noise 66
non-destructive testing 82
non-renewable energy sources 24, 129
NOR gates 227, 229
normal 53, 190, 197
north pole 88
northern lights 89
NOT gates 227, 228
notes, musical 212
nuclear bombardment 256, 257
nuclear decay 238-240
nuclear energy 19, 42
nuclear equations 240
nuclear explosions 74
nuclear fission 184
nuclear fuel 24
nuclear fusion 92, 186
nuclear power 24-26
nuclear radiation 71, 79-82
nuclear reactors 184, 256
nuclear waste 74, 185, 263
nuclei 71
nucleon number 240
numerical data 268

O

ohms 20
oil-fired power stations 33
oil spillages 24
opposing force 111

optical fibres 54, 198, 248
optical telescopes 98, 99
OR gates 226, 228, 229
orbits 106
ordered variables 268
oscilloscopes 143, 214, 245
outputs 219, 228-230
oxygen 249, 257
oxyhaemoglobin 249
ozone layer 58

P

P waves 67
palliative care 262
paper rollers 134
parabola 159
parachutes 113
parallel circuits 38, 145, 224
particle accelerators 245, 257
particles 238, 239, 260
passive safety features 119
passive solar heating 29
path difference 207
payback time 17
penetration 72, 73, 79
PET scanning 257, 261, 262
petrol 129
phase 207
phosphorescent screens 245
photocells 28
photocopiers 137
photographs 204
pitch 211, 212
pivots 161
placebo 266
plane mirror 191
planets 86
plugs 147
plumb line 162
plutonium 184
point of suspension 162
polarisation 208
polaroid sunglasses 208
politics 5
pollution 129
positive charge 133
positive correlation 269
positron/electron annihilation 260, 261
positrons 238, 241, 257
potential difference 140, 144, 145, 221, 223, 251
potential dividers 219, 220
potential energy 122, 124
power 46, 127, 128, 181, 254
power loss 182
power ratings 47, 150
power stations 42, 43
power surges 89
precision 268
predictions 1
prenatal scanning 62, 216
pressure 235, 236
primary coil on a transformer 180
principle of the conservation of energy 19
printers, laserjet 137
probes 97
processing signals 66
projectiles 159

Index

protective suits 80
proton enrichment 257
proton number 257
protons 71, 72, 239
protostar 92
pulse oximeters 249
pumped storage 33
pylons 43
Pythagoras' theorem 156

Q

quality control 215
quality of musical notes 212
quality of signals 66
quarks 241

R

radiation 15, 16, 71, 238,
254, 261, 262
radiation exposure 262
radiation sickness 79
radiators 14
radio 60
radio telescopes 99
radio waves 50, 57, 60
radioactive decay 76
radioactive elements 26
radioactive isotopes 71-73, 76,
81, 184, 239, 257, 261
radioactive waste 27
radioactivity 239
radiocarbon dating 83-84
radiographers 61
radiotherapy 262, 263
radiotracers 261
radius 166
radon gas 74
rainbows 197, 200
random error 267
range 267
rate of energy transfer 127, 254
rate of flow of electrons 244
ray diagrams 191, 193,
194, 201-204
reaction forces 111
real images 189, 193,
202, 204
rectification 221, 222
red giant 92
red-shift 93
reflection 52, 53, 62, 63,
189-191, 193,
194, 210
refraction 52-54, 68, 189,
190, 200-203, 210
refractive index 196-198
relative charge 241, 244
relative speed 154
relays 230
reliability 3
renewable energy resources 25
reprocessing 27
reproducible results 2
repulsion 133
reputation 5
Residual Current Circuit Breaker
(RCCB) 147
resistance 37-39, 140,
144, 145, 149
resistance forces 113
resistors 219, 226, 230

resolution 99
resultant force 109, 110
resultant moment 164
resultant velocity 155, 156
retina 189
road surface 115
rocks 74
roller coasters 125

S

S waves 67
Sankey diagrams 20
satellite television 60
satellites 167
Saturn 86
sawtooth waveforms 212
scalar quantities 154
scattergrams 269
seat belts 117, 118
secondary coil on a transformer
180
seesaws 163
seismic waves 67, 68
seismographs 67, 68
semiconductors 221
sensitivity of an instrument 268
series circuits 144
shielding against radiation 238
shocks, electric 132, 148
shooting stars 86
silicon 221
sine waves 212
skin cancer 58
skydivers 113
sliding friction 112
slinky springs 206
slip rings 179
smoke detectors 81
smoothing circuits 224
Snell's law 196-198
social factors 7
soft (magnetically) 173
solar cells 28, 29
solar flares 89
solar panels 29
Solar System 90, 96
solenoid 173
sound 19, 207, 210-212
South Pole 88
space telescopes 98
spacecraft 96
spanners 161
sparks 133, 134
specific heat capacity 8, 9
specific latent heat 11
speed 102-104, 154
speed limits 115
speed of light 196, 197
speed of sound 214
speedometers 65
split-ring commutator 176
splitting uranium 256
spring balance 106
spying 167
stability 71
curve 239
of nuclei 239
stable isotopes 71, 239
stars 86, 92
static electricity 132, 134, 136
static friction 112
statistics 5

step-down transformers 43, 180
step-up transformers 43, 180
sterilising medical instruments 82
stopping distances 115, 124
strokes 251
Sun 25, 58, 89
supernovas 92
surface area 16
survival blankets 16
symbols 141
systematic errors 267

T

telephone wires 65
telescopes 98
television 245
transmission of 60
temperature 38, 39
of gases 233, 234
sensors 39, 220
tension 166
terminal speed 113
test circuit 140
testing drugs 266
theories 1
thermal energy 19
thermal imaging cameras 17
thermal neutrons 256
thermistors 39, 220
thermograms 17
thermometers 65
thick lead shielding 73, 238
thickness control 81
thinking distances 115
tidal barrages 31
tidal power 25
timebase 142
timing circuits 224
total internal reflection 54, 197,
248, 249
tracers 81, 256
transfer of energy 46
transformer equation 181
transformers 43, 180, 182
transistors 226
transverse waves 49, 67, 208
truth tables 226-228
tsunami waves 68
tumours 261
turbines 40
turning forces 161

U

ultrasound 62, 213-216
imaging 213, 216
ultraviolet 50, 57, 58
unanswered questions 6
unbalanced forces 110
Universe 90, 94
unmanned probes 97
unstable isotopes 71, 74
unstable objects 164
up-quarks 241
upright images 189
uranium 26, 83
uranium-235 184, 256
uranium-236 256
Uranus 86
useful energy 20, 22
UV radiation 50

V

vacuum 211
validity 4
variable resistors 38, 140,
141, 219
variables 266, 268, 269
vector quantities 154
vector sums 159
Vela pulsar 99
velocity 102, 103, 154,
155, 157, 166
velocity-time graphs 105
Venus 86
vertex - centre of mirror 192
vertical motion 159
vibrations 8, 13, 208, 210,
212, 213, 215
virtual images 189, 191,
193, 194, 202-204
visible light 50, 57, 58
visibility 115
voltage 37-40, 43, 133,
140, 149, 151, 219,
223, 230
voltage-current graphs 38, 141
voltmeters 140, 144, 145
volume of gases 235, 236

W

wasted energy 20
watches 65
watts 46, 127
wave generators (converters) 31
wave power 31
wave speed 51
waveforms 212
wavelength 49-51, 200,
207, 212
waves 49-52, 60-63,
206, 207, 213
uses of 60-63
weather 25
weather satellites 168
weight 106
white dwarf 92
white light 200
wind turbines 28
wiring a plug 147
work done 121, 122, 124, 127

X

X-plates and Y-plates 244
X-ray telescopes 99
X-rays 50, 61, 89, 245,
262, 263

Y

Y-plates and X-plates 244
Young's double-slit experiment
207

Z

zebra 189
zero resultant force 109